Co-Evolution of Nature and Society

Jens Jetzkowitz

Co-Evolution of Nature and Society

Foundations for Interdisciplinary Sustainability Studies

Jens Jetzkowitz
Helmut Schmidt University
Hamburg, Germany

ISBN 978-3-319-96651-9   ISBN 978-3-319-96652-6   (eBook)
https://doi.org/10.1007/978-3-319-96652-6

Library of Congress Control Number: 2018948721

Cover credit: memblo/gettyimages
Cover design: Fatima Jamadar

This Palgrave Macmillan imprint is published by the registered company Springer Nature Switzerland AG
The registered company address is: Gewerbestrasse 11, 6330 Cham, Switzerland

Preface

This book presents a conceptual underpinning for a science of coevolution that, I hope, can soon rightfully claim to produce knowledge for a sustainable development of society. My considerations are the result of more than 15 years of research practice in environmental research and sustainability science. But even before, environmental protection and nature conservation were close to my heart. One of my earliest memories is from the late 1970s, when I was collecting signatures at school for the abolition of commercial whale-hunting. Back then, I found it inexplicable, even a bit embarrassing how some of my teachers refused to sign the petition. Perhaps, it was this particular experience that taught me very early in my life that environmental activism alone—as important and valid as it is—will never be enough to maintain a diverse and complex web of life on earth. To save nature, you have to focus on the human societies, you have to study, and understand them objectively; we can only find ways towards sustainable development on the basis of what we know about the relationships of societies with their natural environments. To do so, it is important to outline a conceptual research framework that allows us to correlate problem definitions from the social as well as the natural sciences, and to examine the way natural and social phenomena interact.

The book is a condensed and abridged version of my habilitation thesis on social science knowledge for the discourse of sustainability, which I submitted at the Helmut Schmidt University/University of the Bundeswehr Hamburg in April 2014. I am publishing my thoughts on the methodological foundations of a science of coevolution in book format, to hopefully encourage a new kind of thinking about fundamental tenets of sustainability research. I would be delighted if it contributes in any way to this end. Recent experiences during my work in science-policy interfaces made me realize again the importance of, and indeed the need for such a new approach.

Hamburg, Germany   Jens Jetzkowitz

Acknowledgements

There are so many people to whom I owe a debt of gratitude, among them Udo Kelle (Helmut Schmidt University Hamburg) and Paul Burger (University of Basel) for their support during the habilitation procedure, as well as Stefan Brunzel (University of Applied Science Erfurt) for his many profound and inspiring discussion impulses. This book is dedicated to my wife Kirsten Dietrich, who never let me lose my bearings, textual, or otherwise.

Contents

Introduction

When we look at contemporary society and its relationship to nature, we have come to believe that we know what went wrong: Modern technologies and the organizational procedures of capitalist economy providing resources for markets have had a disastrous impact on ecological systems. But this knowledge is not (or not effectually) acted upon, which is why sustainable development does not happen. Thus, the general thrust of public debates about the protection of the environment and nature.

Yet as sympathetic as we may feel toward the moral integrity of the environmental movement and of environmental politics, we should still critically (and self-critically) examine its wishful thinking whenever social change is talked about. The subject of sustainable development is neither "humanity" nor a person who can be led back on the right path by simple suggestions of how to act right. Already in 1986, Niklas Luhmann pointed out that moral admonitions and reprimands "are not hard to supply. All that is necessary is to consume fewer resources, burn off less waste gas in the air, produce fewer children" (Luhmann 1989, 133). But addressed to the real subject of sustainable development—society—they result in little actual change. And yet, for the last thirty years, nobody seems to have been seriously bothered by this situation. Measures were taken to stimulate ecological change in society, measures that seem painfully naive in light of the insights from the social sciences in the tendency of institutions to decouple from the purpose of their origin and to take on a life of its own. Social scientists have observed such tendencies for more than 120 years and have described them as a main characteristic of contemporary global social structures. Is it any wonder that, despite all the efforts of the last 30 years, the global ecological crisis has deepened, the extinction of species increased exponentially, and the turnaround to decrease emissions of gases harmful to the environment never happened?

At present, this view of our current knowledge base seems at least to be called into question. At international intersections of science and politics—the Intergovernmental Panel on Climate Change (IPCC), the Intergovernmental Science-Policy Platform on Biodiversity and Ecosystem Services (IPBES)—as well as in science networks for global change, people start noticing that barely any scholars from the social sciences and the humanities participate in debates and discussions. The sole exception is economics. To observe and analyze the productivity of nature and the environment is part of an economist's job (cf. e.g. Cortekar et al. 2006), which is why economists have always participated in sustainability discourse, developing their own suggestions on how to end the ecological crisis (cf. Jetzkowitz 2011). But the unquestioned trust once put in an economic rationality seems to have been exhausted. There is a demand for scholars from all the social sciences, to help search for ways to achieve sustainable development. And once they seriously engage with questions of economic management, once they help develop solutions to the conundrum of how, on a global level, the well-being of all can be guaranteed and at the same time dynamic adjustments to changing environmental condition can be made—without destroying nature and the environment—then they will be heard in a scientific community that up until now has been dominated by the natural sciences.

But the perspective of the social sciences in sustainability discourse cannot simply be strengthened by lending an ear to those who call themselves "social scientists". Opening up prevailing discourse coalitions to a new group of stakeholders does not automatically guarantee that problems of sustainable development can be addressed successfully. Rather, it is essential that the community makes room for a new—or at least not yet fully considered—rationality. With this book, I mean to contribute to this kind of change. I propose a conceptual underpinning to adequately conceive of the issues of natural and social sciences alike and to scientifically examine their interdependencies.

I envision a science of coevolution shaped by various sources: ecosystem research, system theory, and the concept of coevolution first developed by Ehrlich and Raven (cf. Ehrlich and Raven 1964). More specifically, I have been influenced by Richard B. Norgaard's book *Development Betrayed: The End of Progress and a Coevolutionary Revisioning of the Future* (Norgaard 1994) and the thoughts and ideas of Marina Fischer-Kowalski and Helga Weisz (cf. Fischer-Kowalski and Weisz 1999) as well as Elinor Ostrom (cf. Ostrom 2009).

Of special importance to my approach is an outline on knowledge production for sustainable social development, the result of a cooperation between Hans-Joachim Schellnhuber (2001) and Helga Nowotny (2001). Combining Schellnhuber's otherwise rather technocratic concepts for earth system research with Nowotny's ideas of a changing relationship between society and science, they describe what an international discourse coalition for "sustainability science" (cf. Kates et al. 2001) could look like. A new *scientific* perspective emerges for the search for reliable knowledge, a perspective that is not interested in creating sustainable social development by simply putting knowledge into practice, or in supposedly practice-oriented research tailored to the interests of stakeholders. Searching for the one miraculous cure, or the one determining the factor to reorient social development toward sustainability should not be the goal of the scientific engagement for sustainable development. Instead, the overall goal should be the explicit research of interdependencies and coevolutionary contexts.

For such a science of coevolution to be able to reliably inform decisions about future social developments in regards to aspects of sustainability, I think it is crucial to clarify what we actually do when we produce knowledge. Any act of cognition, any search for a tenable point of view is an action invariably conducted under conditions that apply to action in general. Knowledge is produced by people enmeshed in sociocultural contexts who developed their individual skills within the context of their personal likes and dislikes. How then can a coevolutionary science be conceptualized under these conditions, a coevolutionary science that generates knowledge of the objective coevolution of nature and society and considers itself as a subjectively coevolving factor in this interdependent dynamic?

Hence, this book is about knowledge as the precondition of sustainability discourse. In the following four chapters, I discuss possibilities and limitations of the systematic production of knowledge about the coevolution of nature and society, knowledge that supports the shaping of social conditions that are built on a premise of sustainability. When discussing this problem, the term "society" invariably takes center stage. For "sustainable development" in fact means the "sustainable development *of society*".[1] Proponents and supporters of sustainable development want precisely this: to instigate specific and purposeful changes in contemporary societies so that they preserve rather than destroy the physical conditions of their existence.

One look at the current knowledge base of sustainability discourse reveals a rather lopsided situation. Hardly any of the insights social scientists have been debating about in the last 150 years (cf., for many others, Elias 1977) found entrance into the field of sustainability studies: What constitutes social change? How can social change be understood and examined? If and under which conditions can social change be purposefully stimulated and oriented toward specific goals? Yet, the answers to these questions are highly relevant to sustainability studies even when they often cannot be easily applied to the shaping of society or used as "techniques". If we take note of how the rules and structures governing societies and social change can be examined and analyzed, we are in a position to objectively assess the possibilities, the difficulties, and the impossibilities to purposefully change society, and avoid at least some

---

[1] The report of the United Nations' "World Commission on Environment and Development" (WCED) from 1987 already puts it like this: "Sustainable development requires that *societies* meet human needs both by increasing productive potential and by ensuring equitable opportunities for all" (WCED 1987, ch. 2, I, 6, emphasis J. J.).

of the monotonous recycling of false assumptions. This applies in particular to the expertise needed to design and conduct research studies, which social sciences also bring to sustainability research, another methodological treasure trove rarely acknowledged outside of the specialized strands of discourse of the social sciences. This should change.

Hence, Chapter 1 explains why scientific knowledge about society is essential for sustainability discourse and why still so little of it has found its way into current debates on sustainability. The chapter starts with a discussion of Marx and Engels' methodological approach to society and history, drawing attention to the problematic relationship of conceptual thinking, future expectation and action, and especially to knowledge as a dynamic feedback process within this relationship. For sustainability discourse, it is crucial to take knowledge as a feedback loop into consideration. Therefore, I examine how sustainability discourse currently addresses societies that need changing within analyses of socio-ecological systems. While there is much talk about society, I show how "society" itself is rarely made an object of research. I discuss why social scientists have failed to tackle this task and why the social sciences as a scientific culture have maintained their distance from sustainability discourse. Finally, I develop the thesis that an interdisciplinary science of coevolution that takes philosophical pragmatism as a point of departure, can help to overcome these barriers. For this scientific approach, it is absolutely essential to acknowledge that any act of cognition and any search for a tenable point of view is an action invariably conducted under conditions that apply to action in general.

Going from there, I ask how we can gain knowledge relevant to sustainability discourse if all cognition is invariably action. Complex issues need to be addressed to answer this question. Thus, I illustrate the problems that sustainability discourse has inherited from an epistemology separating knowing and acting with an exemplary discussion of a classic position, namely, the position of Immanuel Kant. Kant may not the first philosophical reference that comes to mind when dealing with sustainability discourse. But in the history of philosophy, his philosophy comes out of an epochal shift toward subjectivism (and the struggle against it). It is an excellent example to discuss the consequences of the conceptual separation of knowledge and action. Moreover, social scientists ritually invoke Kant. The nineteenth and early twentieth centuries reinterpretation of his philosophy laid the fundaments in the philosophy of science upon which Max Weber built his groundbreaking theory. The current relevance of Kant's conceptual framework may be limited to its role in the history of ideas, yet, it still marks the standard for the type of problems any conceptual framework co-relating knowledge, nature, and action must tackle. The discussion of Kant helps to illustrate both the opportunities and the barriers in thinking that sustainability discourse encounters when conceptually separating knowledge and action. All of this is discussed in detail in Chapter 2.

In Chapter 3, I examine alternative ideas and thought patterns suited to guiding a science of coevolution. The point of departure is not the discussion of abstract concepts but of social practice. Many experts and committees today propose ubiquitous recommendations for shaping the future. From these recommendations I construct a scenario, to better illustrate differences between the previously outlined understanding of reality and the alternative concept I favor. Coevolutionary research is supposed to produce knowledge about a future world that is yet unknown, a world that might take very different paths than we today imagine. My alternative definition of what is reality takes this into consideration. I ask how the science of coevolution identifies its problems, how it produces knowledge, whom it benefits, and how this knowledge is put into practice. Then, I present an outline of the various concepts and ideas, combined into a new understanding of reality. For this I use a discussion partner: actor–network theory, originally developed by Bruno Latour, a former social constructivist, after his turn to philosophical thought. Inspired by the work of Alfred N. Whitehead and Isabelle Stenger, Latour used actor–network theory (ANT) to develop a conceptual framework that claims to have solved the problems at the interface of nature and society. In the social sciences, Latour's framework has garnered much attention. Rather than using ANT, my proposal draws on logic and semiotics to construct an understanding of reality that closely resembles the one we encounter in our everyday lives. Thus in the framework I propose, differences to nonhuman entities are not simply leveled out, as they are in actor–network theory.

In the perspective adopted here, chances for a sustainable social development crucially depend on structures of knowledge and science. While the discussion in the social sciences has been preoccupied with the notorious distinction between knowing and ignorance, little attention has been paid to the category of research or to the question of how people actually find out things in their daily lives. What would happen if we move this process of everyday discovery to the center of our thinking about cognitive and social structures of knowledge? I discuss this question in the context of ideas for a reconfiguration of contemporary knowledge production, taking especially the efforts to establish transdisciplinary science into account. Furthermore, I briefly examine the methodological strategies and the institutional environment of coevolutionary research.

I am not developing my own theory of sustainability. I want to emphasize this; it is why I simply use the term "sustainable development" without further explanation. In my approach "sustainable development" is understood as the topic of a legitimate social discourse, raising questions of ethics and moral values. To discuss and answer these questions is the task of practical discourses and of ethics; it is not a task empirical research can solve. But I am convinced that sustainability discourse as a whole will profit when everyone involved can base their arguments on knowledge that is both reliable and tenable.

References

Cortekar, J., Jasper, J., & Sundmacher, T. (2006). *Die Umwelt in der Geschichte des ökonomischen Denkens*. Marburg: Metropolis-Verlag.

Ehrlich, P. R., & Raven, P. H. (1964). Butterflies and Plants: A Study in Coevolution. *Evolution, 18*, 586–608.

Elias, N. (1977). Zur Grundlegung einer Theorie sozialer Prozesse. *Zeitschrift für Soziologie, 6*(2), 127–149 [also published in English: Elias, 1997: Towards a Theory of Social Processes. *British Journal of Sociology, 48*(3), 355–383].

Fischer-Kowalski, M., & Weisz, H. (1999). Society as Hybrid Between Material and Symbolic Realms. Towards a Theoretical Framework of Society-Nature-Interaction. *Advances in Human Ecology, 8*, 215–251.

Jetzkowitz, J. (2011). Ökosystemdienstleistungen in soziologischer Perspektive. In M. Groß (Ed.), *Handbuch Umweltsoziologie* (pp. 303–324). Wiesbaden: VS Verlag für Sozialwissenschaften.

Kates, R. W., Clark, W. C., Corell, R., Hall, J. M., Jaeger, C. C., Lowe, I., et al. (2001). Sustainability Science. *Science, 292,* 641–642.

Luhmann, N. (1989). *Ecological Communication* (J. Bednarz, Trans.). Chicago: University of Chicago Press [originally published in German: Luhmann, N. (1986). *Ökologische Kommunikation. Kann die moderne Gesellschaft sich auf ökologische Gefährdungen einstellen?* Opladen: Westdeutscher Verlag].

Norgaard, R. B. (1994). *Development Betrayed: The End of Progress and a Coevolutionary Revisioning of the Future.* London: Routledge.

Nowotny, H. (2001). Vom Geschichtenerzählen zur Koevolutionswissenschaft. *GAIA, 10*(4), 262–264.

Ostrom, E. (2009). A General Framework for Analyzing Sustainability of Social-Ecological Systems. *Science, 325,* 419–422.

Schellnhuber, H.-J. (2001). Die Koevolution von Natur, Gesellschaft und Wissenschaft – Eine Dreiecksbeziehung wird kritisch. *GAIA, 10*(4), 258–262.

1

What Is the Problem?

How Knowledge About Society Influences the Development of Society: A Lesson from Historical Materialism

"The bourgeoisie, during its rule of scarce one hundred years, has created more massive and more colossal productive forces than have all preceding generations together. Subjection of Nature's forces to man, machinery, application of chemistry to industry and agriculture, steam-navigation, railways, electric telegraphs, clearing of whole continents for cultivation, canalization of rivers, whole populations conjured out of the ground — what earlier century had even a presentiment that such productive forces slumbered in the lap of social labour?" This is how Karl Marx and Friedrich Engels (1977, 40f.) describe the major innovations of their time in their Communist Manifesto of 1848. The new technologies reflected the radically altered living conditions worldwide, and closely related—as mentioned in the quote—the establishment of the middle-class and the capitalist economy supported by it. Marx and Engels were well aware that the new technologies also profoundly affected the natural foundations necessary for the existence

© The Author(s) 2019
J. Jetzkowitz, *Co-Evolution of Nature and Society*,
https://doi.org/10.1007/978-3-319-96652-6_1

of societies. They saw, too, that changing natural conditions in turn affected social relations[1] even when they had a limited understanding of these interdependencies. Fundamentally problematic is their view of how knowledge takes form and effect in processes of social development. In my view, this leads to a fundamental conceptual error in their theory of social development.[2] Which is why it is worthwhile to study Marx' and Engels' work, for we can learn from their mistake.

Let us briefly look at how Marx and Engels address the issue of internal changes in a society. In every society, they wrote, exist a class of rulers and a class of the oppressed. Both classes try to advance their interests; class struggle shapes the history of human societies. In hindsight, different types of societies can be made out in the course of history. They are each defined by specific conflicts about the modes of production. These conflicts between ruling and oppressed class can only be solved—so Marx and Engels—when new relations of production evolve from the old ones. This dynamic, they claimed, was essentially also true for bourgeois society. They were well aware that bourgeois society brought forth wealth and a degree of individual freedom like never before in history. But the wealth and civil liberties were built on the poverty and misery of an oppressed class, the proletariat, whose existence was mostly shaped by exploitation and alienation. According to what they assumed were regular processes of social development, Marx and Engels expected that "revolution … is imminent" (Marx and Engels 1978, 245, translation J.J.). And with the Manifesto they wrote for the Communist League, they tried to actively participate in those revolutionary processes.

---

[1] It suggested itself to them naturally, for reasons of terminology alone. Their concept of "work", the basic physical engagement of humans with the external world, includes the idea that man, who physically is part of nature, is "acting on the external world and changing it", and thus, Marx wrote (1887, 127), "he at the same time changes his own nature."

[2] The scope and quality of such changes are discussed today as part of the "knowledge society". For the concept of a knowledge society, cf. Stehr (1994) and UNESCO (2005); for an overview of the German debate, cf. Heidenreich (2003) and Engelhardt and Lajetzke (2010).

Today, more than 150 years after the Communist Manifesto was first published, the stated views still inflame the passions: Some dismiss it as a colossal delusion, others praise the brilliance of its conceptual framework and its analyses. Still others are still waiting for the proletariat to rise against its oppressor, and they come up with various reasons for why the revolution has not yet taken place. But for the most part, debates about whether Marx and Engels got it right or wrong, quickly end in an ideological stalemate. I find it more constructive to look at what is pretty much common sense, namely that nobody today would seriously claim that the works by Marx and Engels and especially the Manifesto of the Communist Party have been without consequences. To put it differently: Their works (as well as the works of other writers of the workers' movement) propagated new ideas and started new processes of thought and communication. Individual works, like the Manifesto and Marx' seminal *Capital: Critique of Political Economy* (1867), have become important signs. They may be contested signs, but they embody a kind of knowledge that has given people the power to act (indeed with a particular independence as to whether the purported views are correct or questionable).

It is obvious that Marx and Engels, in their conceptual framework, did not sufficiently consider to what extent knowledge may influence the development of societies. They scientifically analyzed society but could not imagine the immense social transformations resulting from the institutionalization of scientific rationality. Of course, they would not have denied that knowledge can considerably influence the shaping of social conditions. Knowledge, in their view, is a result of social conditions. It can thus become itself an object, used to ascertain whether one's own actions are correct. Marx and Engels did not give enough attention to this second aspect, though. They developed their conceptual framework out of a critique of German idealism, but did not see how their new knowledge about social developments could bring people to correct their behavior in such far-reaching and comprehensive ways that eventually structural laws were invalidated or transformed. Instead, they adhered to a rigidly determinist framework for their analysis of social history. They never understood that with Darwin's theory of evolution a new conceptual world had emerged that made it possible to describe social development neither as a determinist process nor as a wholly random one.[3]

For sustainability discourse, the idea of an open but not random future is constitutive. One cannot participate in sustainability discourse from a position at the end of history. For implicit in the concept of sustainable development is the hope that societies will undergo fundamental structural changes and find knowledge-based solutions to the problems of their existence—including their internal contradictions. The concept is based upon two assumptions (for social theory indeed ridden with prerequisites), namely that social developments can be goal-oriented, and that knowledge is essential for the shaping of societies. These assumptions lie at the root of other discourses of social practice, as well, whenever members of a society communicate about ethical problems of life—be it tax evasion or assisted dying. But sustainability discourse differs from these in at least one essential aspect: The concern for sustainable development is built on the insight that all life and action is bound to material conditions, and it consequently aims to clarify how current interdependencies between society and nature can be organized in a way as to not jeopardize the existence of society in the future. Asking for a way to bring about nondestructive interdependencies between nature and society dramatically intensifies the need for reliable knowledge, a problem other practical discourses struggle with as well. For another problem is added to the problem of how to make predictions about future events: How can research in the natural sciences and the social sciences be correlated, to develop reliable knowledge about how societies can be shaped and about the future consequences of those changes?

This last point is far from trivial. The production of knowledge usually differentiates—often with good reason—between the research subjects of natural sciences and social sciences. Up until now the relevant academic discourse is mostly characterized by conceptual and methodological uncertainties. These intensify when statements are made about

---

[3]The same is true for how Marx and Engels saw Charles Darwin's *On the Origin of Species*. They both appreciated the disenchanting effect of Darwin's theory of descent. But his conceptual and methodological arguments remained for the most part incomprehensible to them. Cf. Lucas (1964), Liedman (1998).

interdependencies that are defining the limits and possibilities of future options to shape society.

For what will in the future be considered an event that is likely or certain to occur, depends largely upon what kind of structural laws are taken as the basis of social development. Let us, for example assume I consider class struggles an objective law of social development that is true at any given time or place. Then it is plausible that I will wait for the proletarian revolution, not only at the end of the nineteenth century,[4] but even still in the twenty-second century. Statements about the necessary ecological embedding of contemporary societies depend just as much on conceptual preconditions. Which processes are classified as "natural" or "worthy of preservation", for instance, is defined not least by how nature and society are conceptualized and how they are differentiated from each other.

And already we are in the middle of a discussion about fundamental methodological questions of a science of coevolution. But first, we can conclude what we have learned from the discussion of Marx and Engels' theoretical framework: Theories and conceptual frameworks can only adequately explain social change, when they account for the fact that the knowledge they produced has itself become an enabling factor of change, and when they address their own transformation.

The Society That Needs Changing in Sustainability Discourse

"Sustainability discourse" can be defined as all the efforts to preserve the conditions that make it possible for societies to exist in the future. These efforts can include verbally articulated reflections such as moral appeals, ethical concepts, and research results. They also include actions that directly or indirectly aim at securing the future of societies. For any

---

[4]Engels, for example, expected to see the revolution happen in the near future even shortly before his death. As late as 1892, he wrote to Laura Lafarque: "Of course, the next revolution which is preparing in Germany with a consistency and steadiness unequaled anywhere else, would come of itself in time, say 1898-1904" (Engels 1979, 545, translation Lawrence and Wishart 2010).

action is always embedded in social contexts and has social meaning, even when it is not directly related to other people but is performed as a solitary act. For example, if someone throws their recyclable paper in the bin for nonrecyclable waste and not in the appropriate paper recycling bin, they—however obliquely and without overt intent—tell their fellow citizens that they do not believe recycling to be a feasible sustainability strategy.

My own definition follows the concept of discourse, as it is used in classical pragmatism and poststructuralism. Consequently, I am *not* establishing "discourse" and "action" as opposite terms, and discourse as a concept is not used to establish reflections and debates as subordinated to action (cf. e.g. Habermas 1981a, b). Quite to the contrary, both terms relate to the same subject, namely social life. But they look at it from a different point of views: From the point of view of action, society is seen as a conglomerate of purposeful activities; from the point of view of discourse, society is a conglomerate of various ways of making sense of reality. The terms thus imply each other. Accordingly, a discourse includes all human manifestations that can convey meaning.

The differences between discourses can be seen in their specific, clearly distinct themes. Sustainability discourse is not identical with nature discourse or environmental protection discourse. And yet, there is a certain overlap between these three discourses. How "nature" is differentiated from "society" is important to sustainability discourse as well as nature discourse. But the relevant texts, actions, and images are linked to different elements in the two discourses. Here, the concept 'discourse' establishes a basic order. In its name public debates can be observed with specific subjects and thematic correlations in mind, traditions of lines of thought can be made transparent, and the developments of meaning and knowledge structures can be traced. And obviously research need not be limited to evident conclusions about a subject, but can focus on their correlations—whether with manifest or latent effects—with meaningful statements or practices in order to establish and generate knowledge about the discursive universe that is, that which can be debated (cf. Boole 1854, 42ff.).

Discourses are variable over the course of time. With the institutionalization of a scientific rationality, social structures emerged, specializing in the production of new knowledge. But under the present conditions it is not only knowledge that increases. With the differentiation of special structures for knowledge production, the criteria for ordering knowledge undergo constant change as well. Discourse analysis itself contributes to increasing knowledge, as well as to revising the criteria we use to order it. For when we ask how distinct sociohistorical conditions lead to the development of and changes in thought patterns, we also raise the question, at least implicitly, of the legitimacy of discourses and generate as a result pressure of justification. To take a stand within a concrete discourse, or to make a statement about this discourse are thus not fundamentally different actions, which are per se equipped with different potentials of thought.

In the following, I first look at sustainability discourse from various perspectives, to find out what it is really all about. Then, I describe in some detail several common exemplary models of goal-oriented social change, all of which can be found in some form or other in sustainability discourse. Subsequently, I discuss how systematically produced knowledge about society can significantly improve the bases for discussion and decision-making in sustainability discourse.

Sustainability Discourse in Historical Perspective: Problem, Concept, Effects

Sustainability discourse combines all efforts to regulate social life in ways that do not jeopardize the existence and the potential development of people in the future. Why such a discourse emerged and how it evolved over time can be explained by looking closely at the history of its central problem, of the concept itself and of its effects.

On first glance, the history of the central problem is quickly told. The problem of sustainable development is a fundamental problem of society, and as fundamentals are usually trivial and not complicated to explain, there is not much to discuss. The existential problem of self-preservation and securing one's continued existence is ubiquitous and holds true for all societies. People have always dealt with basic questions like what to eat the next day, how to protect themselves and their offspring against bad weather, dangerous animals, or human enemies, or how to secure social cohesion in their group. More generally, the provision of resources, the defence against threats from outside, and the management of internal conflicts are the fundamental problems of historical as much as contemporary societies. So that not every generation has to start from scratch when solving these problems, the transfer of experiences and knowledge needs to be regulated, to increase society's chances of a continued existence in the future. Furthermore, the solution—at least partial or limited to a specific time period—of either one if those problems may impact other fields, and nonlinear relationships and interdependencies within societies are to be expected. This is a long-known fact, as is as the embeddedness of societies within natural circumstances.

"Long-known" here means that it *could* have been long known. Historical research about the relationship between society and environment shows us clearly how closely the exploitation of natural resources is related to institutional innovations, which may entail even further institutional change or tensions within society, and perhaps even destructive conflicts (cf. e.g. Radkau 2008). In social theory, too, the interrelatedness of structures within and outside of society is not a big secret.[5] But those insights were never given their due attention. The exploitation of nature continued; when valid concerns were raised, they have so far not lead to any kind of fundamental change. Only with the increasing doubts about the future sustainability of the Western way of life and of an economy geared toward constant growth, is knowledge about the interrelatedness of internal and external structures of society no longer deemed "old-fashioned" and "irrelevant". It appears in a new light, and the question of how this came about, directly leads us from a problem-oriented perspective to a view on the concept.

With that said, on to conceptual history. The discourse of social sustainability cannot be separated from the concept of "sustainable development". It was introduced in the report "Our Common Future",

---

[5]Marx already noted this prominently in *Capital*, cf. fn. 2. Cf. also on Marx, Dickens (2004, 1–28), as well as Groß (2001, 33ff.), who referred also to other classic sociologists.

written by the United Nations' World Commission on Environment and Development and published in 1987. To better understand the concept, several historical and systematic explanations are necessary.

The report of the Brundtland Commission (named after its chairman, Gro Harlem Brundtland) is based on the observation that while industrial societies have successfully secured their present livelihood, they neglect to consider their own future sustainability, as well as the sustainability of the so-called developing countries. Their economic growth and wealth are dependent on exploitation and the acceptance, if grudgingly, of the destruction of ecosystems. In the long run, this cannot work. Which is why, so the report, we need to seriously consider changing the principles by which humans all over the world live and operate economically. There has to be a global perspective for the future sustainability of human societies if we want to stop the destructive impact of the worldwide expansion of modern lifestyles and modern economic systems.

The introduction of the concept of "sustainable development" is the initial step to start a reflective discourse about how to secure the future of human societies. "Sustainable development," it says in the Brundtland report, "is development that meets the needs of the present without compromising the ability of future generations to meet their own needs" (WCED 1987, Chapter 2, 1). Already in the next sentence, this idea is defined more specifically, introducing "limitations" as another key term. Still, the definition of what exactly constitutes "sustainable development" is left open to interpretation. With good reasons, at first rather practical ones. The Brundtland report's goal was to present a broad consensus (cf. Radkau 2011, 536–579; Grober 2010, 249–268; Weizsäcker 1994, 113–128). It was mandated by the UN, an institution with neither the power nor the authority to turn programmatic statements into actual policy. But there is also an epistemological reason: Leaving the concept open to interpretation takes into account that all knowledge depends on one's point of view, also statements about future events are always to an extent unreliable. Who can, after all, know the possible needs of future generations? It makes sense, therefore, to conceptualize "sustainable development" not as a descriptive term but a *regulative idea*. The scope and effectiveness of this concept can be shown in a cursory comparison to its predecessor, the idea of a progressive society that can overcome its given obstacles, and in a short outline of how this change of thought patterns came about.

Sustainability, seen as a regulative idea, is a functional requirement of all social development. There can only be a meaningful discussion about changes, goals, and developments in a society when the society in question will continue, i.e., when it is supposed to have a future. According to Kant, a regulative idea, "does not give us any information respecting the constitution of an object, it merely indicates how, under the guidance of the idea, we ought to investigate the constitution and the relations of objects in the world of experience" (Kant 1899, 496/A671). Thus, we cannot help but presume that struggles for social change will lead to sustainable development, and then start a debate about what is conducive to preserving the options of future societies to shape their fate, and what jeopardizes the future of society. To seek for nonsustainable development is not a reasonable position.

The idea of a progressive society functions differently in that it sets itself apart from other societies, which are deemed "backward", "underdeveloped", and "outdated". In a progressive society things always get better. To make this possible, society overcomes internal and external obstacles and removes constraints. Not everyone in society has to accept this. People, for example, who feel at home in a so-called backward society, will likely be skeptical about the demands and goals of progress.

The idea of a progressive society that overcomes given limitations originates in a worldview deeply rooted in Greco-Roman philosophy and the Judeo-Christian religious tradition.[6] During the seventeenth and eighteenth centuries, it was developed as part of Enlightenment philosophy (cf. Gay 1967/70; Outram 1995; Porter 2001). It was a time that revered reason and science, criticized—even publicly—church and religion, and questioned the legitimacy of monarchy and feudalism. Between 1750 and 1840 major technological upheavals added to this,

---

[6]To avoid a well-known misunderstanding (cf. e.g. White 1967), I would like to point out that to portray the Judeo-Christian religious tradition as the one root cause of the modern ecological crises is an undue simplification.

a process we today know as the "industrial revolution". A new social order not only seemed possible but necessary. In the interconnectedness of all these elements we can see the first contours of a modernity directed toward progress, complete with its control center, the systematic production of knowledge. The internal structure of society changed, while its external boundaries steadily expanded. Spatial distances could be covered in increasingly less time and with more comfort. After the expansion of train and shipping traffic came the automobile and airplane traffic. Rocket technology-enabled humanity to put astronauts in space. It was an on-going success story. But not everybody welcomed these changes and considered them unreservedly as improvements of society (cf. Dickens 2004, 1–28; Linse 1986; Renn 1985; Sieferle 1984).

The new mechanical forms of production had an unprecedented impact on the natural world. Rural exodus and urbanization were the consequences, people migrated to the cities where most of the industrial workforce lived under appalling conditions. Soon voices critical of the status of civilization and of the Enlightenment protested against these and other negative implications of the new social order. But for a long time, these voices were just an off-key note in the otherwise perfect harmony of the ideology of enlightenment and progress. Neither the romanticists nor their successors in the various reform or alternative movements were able to fundamentally disrupt the alliance of progressive optimism. The idea of a progressive society, able to resolve obstacles and limitations, proved to be extraordinarily stable. It survived the horrors of European colonialism, two world wars, genocides and the dread of nuclear armament. It had become an ubiquitous experience, after all, that scientific knowledge, implemented in technology, led to civilizational progress. During World War II, for example, the agricultural usage of artificial fertilizers and pesticides initially made up for the shortage of agricultural workers. Afterwards, it was possible to newly structure the social conditions of existence, for it had become clear that feeding the population of industrial societies could be ensured by using technology and chemicals, with a minimum of human labor.

The idea of progress and the related idea of the emancipation of human society from nature remained unchallenged until the publication of Rachel Carson's *The Silent Spring* in 1962 (Carson 1962; cf. Lear 1997; Kroll 2001; Theobald 2003, 16f.; Nerlich 2003). Carson wrote about the side effects of the large-scale use of pesticides in industrialized agricultural. The book had an unusual resonance. The facts Carson reported are certainly outrageous, even from today's perspective. But the huge stir the book created is no longer comprehensible today. And yet, it was this book that cast suspicion on the blessings of civilizational progress, with its claim that they poisoned society in the medium-to-long run. Even more, it undermined the trust in science, industry, and politics, which had created these blessings and continued to sing their praises. Interestingly, it was not the longings for an alternative, holistic lifestyle which made Carson's argument so effective. Her arguments had such an impact because they were concluded scientifically and based on the specialist reports and expert opinion. What is more, the public debate about the book clearly showed the vested interests of experts from the US Department of Agriculture and their scientific advisory boards. Subsequently, a sense of unease, even fear, became widespread among influential groups in Western industrial societies. They were not afraid because politicians were publicly lying or denied the facts Carson had presented. Unease emerged because all at once everybody could see the inherent limits of technological progress. Suddenly, it became noticeable that the foundations of these achievements—the mechanist and monocausal ways of thinking—were only valid within certain limits. *The Silent Spring* and its aftermath was a first indication that these foundations would ultimately not be sufficient to get a grip on the world.

But can we ever get a grip on the world? Human societies are after all embedded in contexts they did not create themselves. Who can claim to fully understand the impact of human interferences into these contexts? Doubts had been raised, and from now on the environmental movements in the United States and other Western industrial societies could be sure of public attention whenever they pointed out the problematic relationship of nature and society.

The path from these first doubts of the idea of progress all the way to the concept of "sustainable development" was long and rough, without clear points of reference. There were only signs that indicated that the model of a progressive society, ever-growing and ever overcoming obstacles, could not be transferred to the entire world: Animals and plants were going extinct, the average temperature of earth's atmosphere and in the oceans was rising abnormally, and the world's population was steadily on the increase. Also, there was the question of resources: How to feed a steadily growing population, how to cover their need for energy? Energy production by nuclear fission became a case in point, illustrating step-by-step the immense costs of progress and growth. These kinds of observations were made public by the media in Western democratic societies where politicians were already met with a growing distrust. Books like Barry Commoner's *The Closing Circle* and the study "The Limits of Growth" by Dennis L. Meadows and his coworkers intensified the general sense of unease (Commoner 1971; Meadows et al. 1972). Civic organizations mobilized financial and personnel support all over the world, to draw attention to the undesired consequences of resource allocation.

All of this led the United Nations to commission the report "Our Common Future", at the initiative of one its member states, Norway. With the concept of "sustainable development" a new idea becomes visible of how to organize societies and their relationship to each other and to the natural environment. The concept clearly—all openness to interpretation notwithstanding—comes down on the side of global solidarity; it puts an end to the idea that one national society's wealth can be built upon the unfettered exploitation of resources.[7] Since then, the discourse about the future care of society not only has a name; it has been established as an independent discourse. The vague term "sustainable

---

[7]This clarity is a result of the idea of sustainability as it was developed in German Forest Economics, drawing especially from the rule that a forest has to be managed in such a way as to allow for its regeneration (cf. von Carlowitz 2000). This particular aspect was not discussed by the Brundtland Commission (cf. Radkau 2011, 552) but with the word "sustainable" a concept limiting haphazard exploitation was introduced into the discussion of developmental perspectives for economy and society. In 1980, "sustainable development" emerged for the first time as a compromise formula in the international environmental movement. From its origins in the political struggles to end the global destruction of ecosystems and to preserve the biosphere, the formula was adopted into the complicated political process of negotiating a balance between the protection of nature and the environment on the one hand, and the fight against poverty and developmental efforts on the other (cf. Radkau 2011, 536–579; Grober 2010, 249–268).

development" is a deliberate choice, to indicate that an alternative concept for the global development of society is needed. No more but no less, either.

As for its history of effects, the concept of sustainable development quickly gained a lot of attention. During the World Conference on Environment and Development in Rio de Janeiro in 1992, the member states of the United Nations adopted as Sustainable Development as a guiding principle. They committed themselves to focus more on the protection of natural resources and to work against the growing gap between poor and rich countries. An action plan was passed, the so-called Agenda 21, to help achieve these goals. As this action plan was implemented in the member states, on a local level and with the wide-spread support of civic organizations, the term "sustainable development" soon was a household word. Sustainability discourse became a broad-based public discourse. This development also impacted science and research. Numerous small and large projects used, and are still using, the means of the rational problem solving to generate knowledge for sustainable development. This research led to the programmatic establishment of *sustainability science* as an academic discipline in 2001 (cf. Kates et al. 2001).

In the intensified scientific discourse on "sustainable development", the vagueness of the concept, as it was put forth in "Our Common Future", was soon criticized. Rarely is it appreciated that the very vagueness of the concept necessitated myriads of arguments, discussions, and attempts at clarifications all of which fundamentally promoted the discourse of sustainability. Diversity and multiple perspectives are not often appreciated in science. By now, the countless definitions of "sustainable development" in the scientific literature[8] are seen as a problem; the phrase "sustainable confusion" (Jüdes 1997, translation J.J.) is making the rounds. The openness of the concept is especially hard to stomach for scientists who are used to dealing with clearly definable phenomena. And

---

[8]Meadows (2000) estimates that there are more than 70 definitions. Since 2000, this number has substantially increased. Murcott (1997) points out the meaninglessness of attempts to come up with an exact number.

it does foster uncertainties which may lead to misunderstandings and which may eventually leave the door wide open for a potential misuse of the term. Everybody seems to define for themselves what "sustainable" means. Monsanto uses "sustainable agriculture" as their advertising slogan[9]; the German EnBW, an energy company that operates nuclear power plants, professes their commitment to "a safe and sustainable energy supply" in their environmental principles.[10] It is no wonder the term seems arbitrary, without any normative appeal, a catch-all phrase to be filled with whatever meaning seems advantageous to whoever is using it.

To remedy this, concepts and theories have been developed, that define and strengthen the normative content of the concept of sustainability. They benefit from the fact that all concepts of human action have a normative content or even are based upon theories of ethics. This is equally true for the formation of economic terms and theories, even when their advocates may publicly present themselves as representatives of a "logic of factual constraint" (Ulrich 2008, 11; translation J.J.) that will eventually provide for the good of all, if not hindered by morality and ethics. But an economy based on a division of labor cannot function effectively without morality; and any theory of economy is well advised to reflect on concepts of socioeconomic development in terms of ethics. After all, concepts, such as "value" or "goods", as well as "preferences" or "efficiency" are closely tied to notions of how people should act (cf. e.g. Sen 1987). These assumptions provide points of departure for philosophical arguments, aiming to clarify what "sustainable development" means and what not. More and more, the idea of inter- and intragenerational justice emerged as the undisputed normative content of "sustainable development". The idea that all human societies are dependent on physical conditions translated into a general commitment to not allow this complex fabric to be irreparably damaged. Finally, all agreed that care for the future is a universal principle.

---

[9]See this Monsanto website: http://www.monsantoglobal.com/global/au/whoweare/Pages/sustainable.aspx; last access on 17 January 2018.

[10]See this website of "Energie Baden-Württemberg AG": https://www.enbw.com/unternehmen/konzern/ueber-uns/umweltschutz/umweltmanagement/; last access on 14 January 2018.

While the blueprints of and approaches toward sustainable development of society may be legion, they are far from arbitrary. They share a common characteristic: the connection to ideas of justice, the preservation of resources and the care for the future. Depending on the standpoint, they may be vague or concrete, but they all mark a goal for society to move toward.

How to Purposefully Change Societies? Sustainable Development Through Radical Break or Continuous Change

The idea that societies can be purposefully directed toward a certain goal has always been part of sustainability discourse. The predominating question facing people then is what they should do to move a society toward sustainable development becomes paramount. It becomes a moral duty to change the present society. Consequently, a normative understanding of "social change"—in current debates encoded as "transformative change"—is cultivated within sustainability discourse.

Guided by the duty to instigate change, certain views emerged in sustainability discourse of how change processes should look like if they were to bring about a sustainable society. These views, which are identified in categories such as "controlling sustained development" or "environmental governance" (cf. as an overview Voß et al. 2007), oscillate between the idea that a sustainable society will only emerge through radical change, and the counter-idea that continued improvement of existing conditions will lead to a desired goal. In the following, I reconstruct two approaches, exemplifying these ideas as extreme types, to explain how the various views of goal-oriented social change function in sustainability discourse.

"Deep ecology" is an example of the view that a sustainable society can only come about through a radical break with our present way of life. It proposes a spiritual-holistic view of the world, opposing any form of dominance over nature. Founded by the Norwegian philosopher Arne Næss, deep ecology is a school of radical thought that is critical of industrial societies' relations to nature. Central to the approach is an essential and holistic kinship between human and nature. As long as humans are perceived to be *in a relation* to their environment, so Næss, humans will be set apart from nature and thus separated from the environment they absolutely need to exist. Whereas Næss sees humans as part of nature and describes them, like any other organism, as "knots in the biospherical net or field of intrinsic relations" (Næss 1973, 95). This is how he defines a relational and consistently holistic perspective: "An intrinsic relation between two things A and B is such that the relation belongs to the definitions of basic constitutions of A and B, so that without the relation, A and B are no longer the same things. The total-field model dissolves not only the man-in-environment concept, but every compact thing-in-milieu concept—except when talking at a superficial or preliminary level of communication" (Næss 1973, 95). To argue consistently holistic then means to see everything intrinsically related to all else. There is no outside and no observer's position. Nothing is without meaning, i.e. without implication for the world. Hence, nature cannot be conceived of as an "environment" (and thus an entirely different category than humanity) but should be seen as a relational world that in all its diverse forms of expression is fundamentally equal to humanity. Næss, who published his seminal essay on deep ecology in 1972/73, at a time when the destruction, poisoning, and pollution of nature was already discussed publicly, speaks of a fundamental biospherical egalitarianism (cf. Næss 1973, 95f.). Combined with the two other core tenets of deep ecology—diversity and coexistence[11]—it constitutes an antiauthoritarian standpoint critical of the imbalance of power in human societies.

Hence, deep ecologists do not aim to stop pollution and the exploitation of resources by implementing new behavioral safeguards and laws. They believe that a fundamental change of our scientific and technical civilization can only be brought about through a voluntary change

---

[11]Næss (1973, 96) describes the two principles like this: "Diversity enhances the potentialities of survival, the chances of new modes of life, the richness of forms. And the so-called struggle of life, and survival of the fittest, should be interpreted in the sense of ability to coexist and corporate in complex relationships, rather than ability to kill, exploit, and suppress. 'Live and let live' is a more powerful ecological principle than 'Either you or me'."

of consciousness. The voluntary nature of this change is essential; to decree change would be counterproductive. Laws and regulations lead to standardization; they only work when relations within society are hierarchical. Using instruments of power to bring about social change is contrary to the egalitarianism of deep ecology. Goals that can be achieved in such a way are only superficial adjustments of the present social order. Air pollution could, for example, be stopped by regulations setting strict exhaust emission limits. But this is no guarantee that when new technology is developed in the future, the integrity of ecosystems will be even considered. Which is why Næss and other deep ecologists aim to radically change patterns of thought and action. Once people have gained insight into ecology, these changes will unfold locally, autonomously, and in decentralized action (cf. Næss 1973, 98). Hence, deep ecology is not considered a science but teachings of wisdom, whose process of revelation is not yet completed. To this end, fundamental principles are developed, the core beliefs of a pluralistic deep ecology movement, which is open to diverse lifestyles and philosophies and will eventually change social conditions.

All of this suggests clearly that while deep ecology urges us to break with our present way of life, it does not advocated a revolution against current social conditions. The break cannot be an collective act but has to occur individually. The underlying model of social change is well known in Western culture; the prototype can be found in the Parable of the Laeven in the New Testament (cf. Gospel of Matthew 13, 33). There we see another eschatological vision of human coexistence, the kingdom of God. It will come—just as, according to deep ecology, the sustainable society will come—steadily and inexorably, as we change social practices in small steps for the better, by a radical shift in thinking and by individual reform. The laeven model does not tell us how this change is organized, or whether it can be purposefully organized at all. And while this may have been acceptable 2000 years ago, when the texts of the New Testament were written, today, in view of the self-made problems of contemporary societies, we need answers to concrete questions: What knowledge supports the development of a sustainable society, and how can we create it? How will new institutions arise from individuals changing their beliefs, and what are we going to do with the old institutions? How can we balance competing goals of different life-styles? In the face of such questions, deep ecologists point toward the importance of new language and altered concepts for a fundamental cultural change (cf. Sitter-Liver 2000). But this only rudimentarily sat-isfies our present need for orientation. Especially from a social scientist's perspective it can be said: It is indeed possible to develop sound, fact-based answers to these questions and put them up for discussion. The distant goal of a sustainable society—and for deep ecology this means, a society co-existing peacefully with nature—should be up for discus-sion, too. Questions about the shape this distant goal should be part of sustainability discourse, as much as questions about the possible conse-quences of these changes.

Whereas a vagueness about questions of social change is intentionally built into the laeven model, its counter model offers clear and precise instructions for a continuous change of social structures toward sustain-ability. Such a process of change presupposes no prior change of con-sciousness. Rather, the plan is to create new institutions that seem so self-evident to us, or so constraining, that people cannot but behave in a sustainable and environmentally conscious manner. The success of these efforts is then best measured against indicators of sustainability. If nec-essary, the strategies can be adjusted. No break with existing social and economic structures is required; the goal is their reform.

It's easy to adopt this position as one's own. When you consider the existing system as reformable you can point to its evident and (seem-ingly) historically unprecedented successes in securing society's live-lihood. Moreover, this position does not make overwhelming moral claims on the individual. Can such a position then really be charac-terized as "extreme"? Today its proponents are seen as representative of the so-called mainstream; they frame their ideas in the terminology of ecological modernization theories or neoclassical environmental eco-nomics. Still, this position marks one end of the spectrum of views within the discourse of sustainable social development. In the tradition of the model of progress outlined above,[12] sustainable development is

---

[12]Cf. p. 19f.

conceptualized as a continuous, differentiable, i.e. as a linear process. No break with existing structures is required, only their modification, trusting that the recipe for the success of modern society will stand this test as well. After all, to not poison our environment and ourselves, and to not build our wealth upon finite resources, is just another aspect of securing our livelihood. The capitalist market economy simply needs to adjust to material flows with renewable energies and to the maintenance of ecosystem services. A key role plays the search for new, more efficient technologies for energy generation, for the production of materials and goods, and for the organization of services. But, so proponents of this view, the responsibility for a successful outcome lies not only with engineers and natural scientists doing basic research. Technologies, they argue, do not emerge by themselves but within socioeconomic contexts, which makes it important to develop social structures in such a way as to improve what already exists and at the same time encourage the innovation of fundamentally new technologies. A big step in the right direction would be the inclusion of ecological follow-up costs of goods and services into the calculation of costs and prices, as it would create economic incentives for an environmentally sound economic management.

But neither ecological modernization theory nor neoclassical environmental economics see this step in the right direction as something that is inevitably going to happen. In general, the idea of determinate, inevitable social change seems to have been relegated to history. It is superseded by the concept of path dependence. It implies that future opportunities for social development are limited by the paths of development taken in the past. The concept achieves plausibility through everyday experience. At least our near future is prestructured largely by past decisions—the decision for a specific career, for marriage to a specific life-partner, against private pension insurance and so on. Subsequently, ecological modernization theory and neoclassical environmental economics both postulate that existing conditions cannot be changed equally well at all times. "There is a time to plant, and a time to pluck up that which is planted," we already read in the Bible (Ecclestiastes 3:2). In the context of path dependence, we can differentiate between stable and unstable phases of development processes, with the assumption that if a development process has moved from an unstable into a stable phase, very little room is left to alter the chosen path.

Ecological modernization theory uses the concept of path dependence to explain why it is not possible to pursue a sustainable society through a radical break with the present growth-oriented economic practice. In a critique of sufficiency strategies developed from a position on economic growth criticism, Joseph Huber, one of the leading exponents of ecological modernization theory, comments on their potential to actually achieve sustainability: "Sufficiency, in the sense of a voluntary or enforced limitation on consumption, has a relatively small savings potential. More importantly, its potential for gaining support and for resonance is much too small. Under the persistent conditions of a worldwide dominance of materialist, utilitarian values, it is pointless to bother too long with the sufficiency approach, at least in terms of politics. Certainly, from a cultural sociology perspective things would be weighted differently, and on the personal level an anticonsumer ethics may contribute to a more fulfilled life. It will surely benefit our salvation to recall the Old Lutheran saying that man shall not live by bread alone. Yet we do live by bread, among many other things, and to eat less bread, as a result of a pious inner life, has no significant impact on the protection of the environment" (Huber 2000, 119; translation J.J.). It is the pronounced emphasis on constraints on development that makes this rejection of sufficiency strategies seem so convincing. One cannot help but mildly smile at the gospel of saving which appears to be the source of these attempts to achieve sufficiency. Huber's polemic tone, at least, suggests as much.

There is a catch though. It is only a hypothesis—an unverifiable hypothesis—that we need to follow the path of a growth-oriented economy until maybe, in a future far, far away, a phase of radical structural instability will come and open up different economic and social options. It may be a plausible hypothesis for particular cases. But to accept it as a general principle shaping our reality misses the point. To be sure, this hypothesis comes with much suggestive power: Once a growth-oriented economic structure has been established as a development path, we are missing—thus the argument—actual, real opportunities to achieve a sustainable society if we are not taking this path but instead engage in critical reflections on the concept of growth and the search for alternative social and economic structures. But with this argument again a determinist concept of social change and societal development enters into sustainability discourse. For when we take the unloved but conciliatory path of a growth-oriented economy and come up with, let's say, economic incentives to not poison the environment and to go easy on natural resources, we simply reaffirm present conditions like in a self-fulfilling prophecy. We neither challenge the conditions declared to be "path-dependent", nor do we become sensitized to opportunities for a truly fundamental change.[13]

To summarize: Both extreme types refer back to basic categories of a theory of social change. They discuss—one explicitly, the other implicitly—freedom and restriction of action, emphasize the break or the continuity with existing social structures, and they look into the near as well as distant future. They discuss society and rely on strong assumptions about the structures of society. But neither makes those assumptions themselves the subject of their research.

Knowledge About Society: A Desideratum in Sustainability Discourse

Is there anything we can know about our society that is not liable to subjective factors? In a widespread classical view, only those insights are deemed objective that are based on the timeless ideas or mandatory, unalterable concepts. All insights based on the experience, however, are classified as "opinions", as they may be influenced by personal

---

[13]Recent sustainability research experiments with the concept of "pathways" (cf. e.g. Geels and Schot 2007), in order to avoid deterministic interpretations. A pathway can be described as an exploratory movement of various actors on interlinked micro, meso, and macro levels. In hindsight, each pathway is reconstructed as the result of, on the one hand, intentionally planned efforts for change and, on the other hand, emergent properties of social processes. Whether this concept can actually provide an anti-deterministic approach in sustainability discourse depends, I think, in large part on its linkage to co-learning strategies for the shaping of the future (cf. e.g. Luederitz et al. 2016; Didham and Ofei-Manu 2015; Dyball et al. 2007).

epistemic goals. Especially in the natural and technical sciences, this view may correspond with old, well-established reflexes and academic self-images. But the classical dichotomy between knowledge and mere opinion has no validity in sustainability research. Today we assume—not least because of research in economic sociology—hat it is essential for all knowledge processes to consider knowledge with the researchers' intensions and values in mind. There is no knowledge without knowing subjects, and they are always children of their time. And, as the problem of sustainable social development is oriented toward care for the future, the classical dichotomy doesn't apply. As the future is unknown, even what seems like certain knowledge may not be so; there is always a caveat about the certainty of future events. It is true that we can proceed on the assumption that not all future events are uncertain. We live in a world structured by laws and rules, after all. They limit the range of events that may or may not happen in the future. Some of these laws are natural laws, but others are behavioral habits shaping our social life. We conceive of them as general regularities and rules; they are the subject of empirical science research, and we can put them up for discussion and thus develop reliable knowledge about society—i.e. knowledge examined and cross-examined in scientific debate.

Growth criticism is one the most prominent positions within sustainability discourse. A discussion of this position may help to illustrate the new impulses that a stronger reference to knowledge about society may give to sustainability discourse. The study *The Limits to Growth*—published as the first Report to the Club of Rome in 1972—prominently highlighted the question whether the key problem in a finite world with limited resources is its orientation toward continuous economic growth. Among those addressing this question, two camps have emerged, each emphasizing different points of criticism.

The economist Joseph Huber is a representative of the moderately critical camp. In his version of the ecological modernization theory, he emphasizes that "growth", as it is presently defined, is not conducive to structural changes toward a sustainable society. Huber introduces the term "industrial ecology" to clearly distance himself from efficiency strategies for sustainable development (cf. summary in Huber 2011, 286f.).

A change toward sustainability, Huber argues, does not depend on whether more or less of the same material is used; rather it depends on whether different kinds of material are used (cf. Huber 1995, 2000). "Industrial ecology" thus means a fundamental change of the material flow in industrial societies. For, so Huber, ecologically consistent material flows allow for qualitative growth. Contrasting his understanding of "qualitative growth" with earlier definitions, he writes: "Earlier views of 'qualitative growth' were sometimes associated with major misapprehensions: the idea of 'selective growth', for example, implied that wealthier countries stop growing, while poorer countries continue to grow. Or that agriculture and industry should no longer grow and instead services be expanded, as the service industry was thought to be per se environmentally sustainable. For a time, structural changes were assumed to have 'environmental gratis effects', following Clark and Fourastié's three-sector model, which has since been deemed misleading. For in reality, the service and knowledge society is a highly industrialized society throughout all the sectors, and it is many times more material- and energy-intensive than even the traditional industrial society, just as the latter in its initial phase was much more material- and energy-intensive than the agricultural society" (Huber 2000, 112; translation J.J.).

Huber wants to see the growth model refer to fundamentally different, ecologically sound industrial metabolisms, but he does not attack its basic foundations. The opposite camp applies a more radical criticism. It emerged mostly in the field of so-called ecological economics, although it must be stated that eco-economics is no unified school of thought, oriented toward one specific paradigm (cf. Costanza et al. 1997, 48ff.). Rather, ecological economics constitutes an alternative branch within economics that emerged as a pluralist movement in the 1980s, intent on bringing together ecology and economics (cf. Røpke 2004, 2005). "Bringing them together *again*," ecological economics will emphasize and point out that the late nineteenth century differentiation of ecology and economics into two specialized sciences was a mistaken development in the scientific system, a mistake that needs to be remedied (cf. Costanza et al. 1997, 19–75). In any case, the credo of ecological economics is this: Economics need to learn from ecology to consider the limitations of human economic activity in a nongrowing physical world. The growth model is seen as unsuitable for the organization of economic processes, as its reliance on the possibility of unlimited growth, so the criticism, does not match the facts of the real world. Herman Daly and his theory of a steady-state economy profoundly shaped this view. According to him, an economy can be in a stable, stationary position "in which the total population and the total stock of physical wealth are maintained constant at some desired levels by a 'minimal' rate of maintenance throughput" (Daly 1973, 152). Other than Huber's industrial ecology, Daly assumes that material and energy flows are limited and nongrowing. However, Daly does not make a case for rigid, stationary economic structures. He differentiates between (quantitative) growth on the one hand, and developments to optimize material and energy flows on the other hand (cf. Daly 1987). Here the two theoretical positions seem to overlap, for example when Daly writes: "Growth is more of the same stuff; development is the same amount of better stuff (or at least different stuff)" (Daly 2008, 1).

Both moderate and radical growth criticism is characterized by its great trust in the natural and technological sciences. Economists cannot verify such claims, and yet ecological economics highlights the physical limitations of economic growth; in the industrial economy, solar and hydrogen technologies, as well as genetic engineering are declared to be integral parts of future societies. Is this trust justified? Without disputing the professional integrity of natural and technological scientists, we still have to assume that in principle, their knowledge, too, is uncertain. For this reason alone it makes no sense to base scenarios of sustainable development on structural laws that representatives of the natural and technological sciences see as decisive for the future, according to the current state of research in their fields. This is not to deny that their knowledge and their assessments are indispensable. But their statements should only be given special importance when they deal with natural and technological subject matter, and not in regard to what constitute desirable or undesirable consequences of scientific and technological innovations.

It should be further noted that economic—and often added demographic—parameters cannot adequately describe how technologies are socially integrated. They are without a doubt significant, and they provide information about specific aspects. But they say nothing about the internal point of view of individuals—what impels people to act, what scares, what convinces them. And yet to be able to understand people's actions, it is essential to know how people themselves define the situations they find themselves in. For people's own definitions influence how, when and where they act, and what means they will use for their actions. As early as 1928 William and Dorothy Thomas summed up this insight in the formula: "If men define situations as real, they are real in their consequences" (Thomas and Thomas 1928, 572). In light of the so-called Thomas theorem, it becomes obvious that knowledge about the structural laws of a society is highly limited and will remain limited as long as the relevant actors' own definitions of their situation are not taken into account.

This has long been known in many fields of sustainability research. But rarely has it been demonstrated quite as clearly as in Loren Lutzenhiser's work on energy efficiency research in the United States (cf. Lutzenhiser 1993; Wilhite and Nørgaard 2004). The research's goal is to help reduce energy consumption, however, it focuses on the physical and technical aspects of buildings and appliances, and energy prices respectively. No special attention is given to social context. It is thus no wonder that all efforts to reduce energy consumption through energy efficiency research are bound to fail. The research does not look at how energy consumption is embedded in people's diverse life praxis, how practices of their everyday life are connected to building infrastructure and household technologies, and how in general, social and cultures factors influence the demand for energy. When lifestyles *are* examined, research is limited to demographic and psychological variables. The fact that people live in networks of social relations is never considered; nor is ever addressed how lifestyles and actions are developed by identification with others who seem similar, or by differentiation from groups perceived as "other". Energy efficiency research can do nothing but state so-called rebound effects on a case by case basis, which comes down to an acknowledgment of its own failure. The principle this research refers back to, had already been described by William Stanley Jevons in 1866. In his book *The Coal Question*, Jevons demonstrated—using the example of coal consumption in England before and after the introduction of James Watt's more energy-efficient steam engine—that higher efficiency does not lead to a decrease in overall energy consumption. Jevons made his argument almost 150 years ago. Do we know more today? If rebound effects are bound to happen anyway, and people can do nothing against them, why then continue with energy efficiency research? Or is the paradoxical causal relation Jevons described not as inevitable as it seems?

Currently, sustainability research does not seem to have the conceptual tools to answer such concrete questions and base scenarios of sustainable social development upon knowledge instead of postulates. There is no methodology available to evaluate the scientific quality of research results and to assess what progress may have been made (cf. Radnitzky 1989, 465ff.). It is no new insight that sustainable social development needs knowledge about the "interfaces", "connections", or "couplings" of physical and sociocultural structures; it's obvious that this kind of knowledge will increase the reflectivity of sustainability discourse. Unclear is how this knowledge can be implemented systematically and adequately in concrete research work. We need to overcome the barriers between scientific disciplines, most notedly between the academic cultures of the humanities and the natural sciences. In other words, we need a new approach that does not replicate the classical difference between idiographic and nomothetic sciences (cf. Windelband 1904). Rather, this new approach should correlate physical and social structural laws, and at the same time demonstrate how knowledge produced through research can change the world.

Hans-Joachim Schellnhuber may have had such an approach in mind when he, assisted by Helga Nowotny (2001), translated *"sustainability studies"* with the German "Koevolutionswissenschaft" in 2001. This is how he defines *coevolutionary science*, namely as "the closely linked and mutually accelerating development of global nature, society, and knowledge—a process that the academic system has to adequately explain, but even more so, a process in which the academic system needs to reinvent itself. Thus coevolutionary science is, nota bene, at once the objective doctrine of human-environment coevolution, and a subjective coevolving factor in this interrelated dynamic itself" (Schellnhuber 2001, 258f.; translation J.J.). However, Schellnhuber's own earth system analysis approach does not live up to this definition. It remains stuck in the kind of thinking that for the most part relies on natural science and technocratic arguments.[14]

And yet system theory and cybernetics seems to be especially suited for inter- and cross-disciplinary approaches where the objects of the natural sciences and the humanities are correlated as dynamic, self-adjusting systems, whenever necessary. "Self-adjusting" does not mean "controlling"; the focus is not specifically on the question of how societies move toward goals, such as, e.g. sustainability. Rather, all spoken and unspoken rules of human life come into focus, as researchers are encouraged to differentiate regular patterns from structural laws, define potentials for change, and assess possible consequences. For system theory and cybernetics to be successfully adopted into sustainability discourse, it seems crucial to overcome views that see nature as a world of necessities and the mind as a realm of freedom, and to stop lamenting the false distinction between certain, unquestionable knowledge and mere opinion (which is the same as uncertain knowledge). Instead, it is important to produce knowledge about society systematically, with adequate methods, and drawing upon the existing body of knowledge, and then evaluate it by means of quality criteria. Obviously, the results of sustainability research will not become more certain when knowledge from the social sciences is taken into consideration. But it provides sustainability discourse with an increased reflexivity when views of the future viability of societies match up with spoken and unspoken rules of human coexistence today.

---

[14]This is true for the large majority of sustainability science approaches, cf. Bettencourt and Kaur (2011).

Why Is It so Hard for the Social Sciences to Participate in Sustainability Discourse?

It is an essential function of social sciences to provide knowledge about observable regularities and structures within societies. In light of the desideratum outlined above, new impulses for sustainability discourse could well be expected from these sciences. After all, the question of how human coexistence can be best organized in the future has always been at the core of social research. Proponents of this view are found at all times among the ancestors of sociology: Adam Ferguson and Friedrich Schleiermacher, Auguste Comte and Karl Marx, Emile Durkheim and Beatrice Webb. Even opponents, who dismiss research oriented toward concrete social goals (or even social reforms), promote their idea of an academic-contemplative science mostly interested in interpreting culture with the argument that society works best when each part deals with its own tasks. This kind of argument, too, puts forth a vision of the future.

The social sciences may refer inherently to the shaping of the future of society. Still, the issue of "sustainable development" was by no means met with comprehensive resonance. Special fields developed in political science, legal studies, psychology, and other social sciences that reflect upon and reconfigure action strategies. But it cannot be said that the social sciences in general made "sustainable development" their issue or that at least a substantial number of social scientists felt the need to position themselves and do research in this field. It is rather the other way around: For the most part, social scientists are simply not interested in sustainability discourse. They leave it to natural and technological scientists, or to economists, to define the areas and margins of development for tomorrow's society.

Why is it such a problem for the social sciences to get seriously involved in sustainability discourse? Is it perhaps due to the fact that there is no clear-cut discipline of social research, but only a many-voiced, multi-perspectivist hodgepodge of social sciences where even the supposed basic definition of "society" does not go unchallenged? Modern corporate marketing would surely consider it a catastrophe, but this situation is, in fact, part of the subject of sociological research. For all of this impact our research object—the name we give it, how we examine it and which goals and purposes we pursue when we research it. To define, to discuss and to research is to act. And to act is to participate in the organization of our social lives. Debates in social sciences are always inherently debates about what we desire, debates about a good, a right life. Hence, I won't deplore the multiple perspectives and many voices within the social sciences, but analyze the strategies of how sociologists react to these structural features of their field. And doing so, I sound out which strategy may match the concerns of sustainability discourse. That this is a type of strategy most notably exemplified by Critical Theory, will come as no surprise to anyone in the field. It will be interesting to reconstruct why sustainability discourse causes a *double bind* situation also for Critical Theory, and to clarify—most importantly—how the double bind can be overcome.

Social Sciences, Diversity of Perspectives and Sustainability Discourse

Too many cooks spoil the broth, goes the saying. One might argue that the many voices and multiple perspectives in the social sciences are a direct result of the researchers' inclusion into their research object. But this doesn't mean that social research necessarily needs to be ambiguous. The history of social sciences offers many different possibilities when dealing with multi-perspectivity and a diversity of voices. Among the various theoretical and conceptual responses, at least three types of strategies can be made out of how social sciences react to sustainability discourse. In a nutshell, they can be defined as "disciplining of knowledge", "skepticism", and "social criticism".

The first type of strategy, the "disciplining of knowledge", is out to defy the perspectivity of all knowledge, to deny it, so to speak, and to do—if only for moment—what other sciences also do, namely, to restrict all the many possibilities of searching for knowledge. In order to give the own field a recognizable and distinctive profile— as say, "environmental sociology", "environmental politics", or

"environmental law"—study and examination regulations with definitive test contents are developed, mandatory reading lists are put
together, and compendiums of classic texts and primers are published.
What this type of strategy does very well is to provide orientation when
the goal is to gather and structure the body of knowledge currently
available about a certain problematic. There can be no objection to this,
especially not if it has been clearly stated that the selection of knowledge
is subjective and historically specific.

Criticism is called for, however, when this subjective selection of
knowledge—which can never be more than a snapshot in time—is
defined as canon. Then the legitimacy of such strategies is at once called
into question. For to canonize knowledge, you need to see yourself as
invested with the power to establish reliable facts and to close the horizons of interpretation, even when an openness to new knowledge and
new problems in the future is far more reasonable. Of course, there are
situations in all human interactions where power needs to be exercised,
even in academic life. But these situations are not about gaining new
knowledge, they are about enforcing certain interests. Those interests
need to be open to debate, and they must stand the test of argument
before they can be acknowledged. If they are shrouded in an aura of
irrefutability, though, they are only covering up the fact that this type of
strategy always operates (and has to operate) arbitrarily.

In view of sustainability discourse this type of strategy offers two
options: Sustainable development could, for one, be defined as a field
in the environment-related subdisciplines of the social sciences. Or it
could be defined as its own subject, to become a subdiscipline in itself,
called "sustainability politics" or even a "sociology of sustainability".
But neither option does justice to the concept of sustainability. Not
because of the fact that the concept is classified or even subclassified,
but because the concept of "sustainable development" suggests (and
encourages) a specific view of social sciences. Sustainable development
is a regulative idea, and research about it necessarily includes questions
of quality assessment. It is a tiresome topic, one that has been virulent
in the discourse of social sciences for over 100 years and has led to huge
controversies. The concept of sustainable development is not indifferent to those controversies; it demands of the researcher to take a specific

standpoint. Anyone researching questions of sustainability has to believe that the knowledge produced by their work will contribute to a *viable* organization of future societies. There may be good reasons for such a standpoint, but it is a standpoint nonetheless, and not a particular field in the research object of the social sciences. Therefore, debate and research of sustainable development cannot be classified according to the logic of academic disciplines and be organized in, say, subdisciplines of the social sciences.

The second type of strategy—skepticism —works fundamentally different than disciplining. It is used across all the social sciences but crops up most notably where "reflexive knowledge" is set against "instrumental knowledge". While others come up with recipes for the shaping of the future world, the skeptics will hold that in the end, all knowledge is about the knowing subject itself, and never about the object of knowledge. These kinds of strategies add nothing fundamentally new to sustainability discourse. For quite some time now, skeptics have found a sympathetic ear in public debates, when they debunk climate, nature and environmental protection efforts as exaggerated scaremongering. Facts, they claim in their proclamations, are only constructs, and one can be skeptical of constructs, after all (cf. e.g. Vahrenholt and Lüning 2012; Lomborg 2001; Reichholf 2002; Krämer and Mackenthun 2001; Maxeiner and Miersch 1999). And of course one can be. But the argument wears thin quickly when there is no constructive follow-up.[15]

Skepticism should not be confused with criticism. It is no wonder that the third type of strategy, represented by Critical Theory, uses a rather different approach. This type of strategy claims to not only understand but to improve social relations. It does not deny that social sciences are embedded within their specific social and historical contexts but, quite to the contrary, makes the social and historical embeddedness its point of departure from where to develop its own position. Critical Theory, for example, aims by definition to contribute to human self-determination. It is never indifferent or indecisive about

---

[15]That it is possible to gain public attention in the role of a fundamental scepticist indicates that a new issue has been established in public discourse (cf. Jetzkowitz 2008, 104–109).

power relations and power structures. Rather, it evaluates social conditions in light of how they may contribute to a future human existence that is free and without constraints. Unlike Kant, freedom of humanity is not considered a given state, but a state to be desired and fought for. Critical theory does not thus divest itself of a metaphysical commitment. It becomes a science with a calling, namely, to accompany all future social development in a way so it may turn into a process of emancipatory learning.

Conceived of like this, the Critical Theory type of strategy is fully in tune with popular ideas in sustainability discourse. For there, this is clear: No natural law determines the development of society. Humans shape society, and humans can change society, especially when social structures result in consequences threatening society's very existence. It is these views, too, that are the driving force behind the engagement of natural scientists, engineers, and economists for alternative paths of development. Similar to Critical Theory, they search for opportunities of change when deciphering current social practice in view of a better—meaning: a sustainable—practice. Social scientists who contribute to these tasks by sharing their knowledge of social structures and processes to build a sustainable society, should be welcomed with open arms.

Sustainability Discourse: A Critical Theory of Society?

Love, at first sight, is not always a viable basis for a relationship. It is no different with sustainability discourse and Critical Theory. For Critical Theory distances itself from all efforts to infer normative expectations from knowledge produced by the natural sciences. Max Horkheimer and Theodor W. Adorno saw the domination of nature—that the natural sciences are an integral part of—as the origin of all aspirations for domination and ultimately, as opposed the freedom and self-determination of humanity (cf. Horkheimer and Adorno 1969). From their point of view, the everyday business of empirical research—explaining observations with the help of concepts, confronting hypotheses with facts—is part of a division of labor characteristic of present-day social structures.

Critical Theory sets itself apart from this kind of research. To be sure, Horkheimer and Adorno concede that the so-called traditional science supported the emancipation of the middle-class. But it also cemented today's power relations with all their inhuman brutality. Critical Theory instead propagates a new organization of society. But their protagonists, namely Adorno and Horkheimer, can only imagine this with the suspension of the present organization of society. Put in slightly simpler terms, Critical Theory poses a fundamental criticism; society as it is organized today is pulled to pieces. "There is no right life in the wrong one," states Adorno's famous maxim-turned-aphorism.

But if nothing is seen as valid and good in present society, what then remains to gauge and check one's own normative standards? Jürgen Habermas, probably the most well known of the second generation of critical theorists, tries to remedy this flaw by distilling from language and action theories a foundation for the critical observation of society. This distillate he defines as "understanding". According to Habermas, reaching understanding is the *telos* of speech. Societies, writes Habermas, need to take care that their members engage in reasoned debates about public issues and determine future developments through participatory decision-making processes. Hence, his version of Critical Theory advocates to make use of the potential of communicative rationality. The criticism of society can now be held to a standard, one that may also hugely enrich sustainability discourse. The exploitation of nature, the rapid waste of resources, and the destruction of ecosystems and living beings can now be described as consequences of an epistemic interest geared toward human domination of nature, consequences that need to be tamed by the means of communicative rationality.

In fact, Habermas' version of Critical Theory has been proposed as a conceptual framework to interpret the environmental movement in the United States (cf. Brulle 2000), as well as an integrative theoretical foundation for the environmental movement itself (cf. Luke and White 1985). His discourse ethics, in particular, are well regarded as a formalistic process designed in general for the constitution of norms sensitive to social justice issues, while freeing capacities for problem solving as it concerns sustainable social development in particular (cf. Brulle 2000; Zierhofer 1994; Dryzek 1987; Ulrich 1986).

Critical voices find fault with Habermas' approach, as it conceives access to the nonhuman, physical world solely through instrumental reason (cf. Scheunemann 2008; Ott 1993; Eckersley 1992): You don't communicate with nature, but use it, says Habermas, thus emphasizing the categorical difference between humanity and nature. This aspect has occasionally been used to dismiss Habermas' formula for solving the problems of social development. Two lines of counterarguments can be differentiated. The first is put forth in particular by proponents of eco-centric ethics. Human predominance over the nonhuman world will just continue, they point out, if we accept that communicative rationality is supposed to control instrumental rationality. They reject discourse ethics as an anthropocentric process, which does not take into account the interests of the physical world (cf. Eckersley 1992). This argument, however, is easily refuted. Habermas designed the discourse ethics process so it is open to all interests, including those that are brought forth on behalf of others. This includes humans who cannot themselves articulate their interests, as well as nonhuman forms of life (cf. Scheuneman 2008, 68–83; Zierhofer 1994, 179 and 183; Eckersley 1994, 143ff.; Ott 1993, 109ff.).

The second line of argument is not as easily refuted. It builds upon the observation that Habermas takes his definition of instrumental rationality from the older Critical Theory, and thus identifies knowledge of nature with the domination and technical control of nature. It does not make sense, so the argument, to claim that the desire to understand nature (and society) must be per se strategical. Why should it be impossible for a social practice to evolve in the course of objectifying research, a social practice that does not have the technical transformation of the physical environment as its goal? Some scientists, after all, describe their research practice as downright communicative and understanding. Perhaps the desire to understand a natural phenomenon, to wish to know what caused it and what are its effects, constitutes in itself an autonomous form of action. Perhaps knowledge processes in the natural sciences can indeed be understood as interaction with nature? Deep ecology taught us that such ideas have their place in sustainability discourse. They have no place in Habermas' modified reframing of Critical Theory.

Habermas sees this line of argument as merely an alleged flaw; moreover, it is not a flaw that has entered his theoretical framework unintentionally. His critical engagement with those who have found fault with the fundamental separation of reason and nature says as much (cf. Habermas 1995, 505–521). The encounter with "nature itself" (*Natur an sich*)—Habermas' term for "a reality existing contingently and independent of us" (Habermas 1995, 510; translation J.J.)—yields neither epistemological nor ethical impulses that would necessitate changes to his theory. Conversely, the impact of a nonobjectifying relationship with nature on forms of solidarity and knowledge structures, so Habermas, can still be interpreted within his theoretical framework: as ethical institutions, on the one hand, enabling us to express sympathy and solidarity with living creatures[16]; as cognitive impulse on the other hand, to reconstruct the evolution of living beings as the *"pre*-history of sociocultural forms of life" (Habermas 1995, 510; translation J.J., emphasis in original).

When Habermas shows us what kind of linguistically mediated reflections are initiated by encounters with, allegedly, "nature itself", he gains two things: Keeping up the separation of reason and nature allows him to retain the basic tenets of Critical Theory. And he is able to argue convincingly against all demands to base his theoretical framework upon a fundament that does not strictly distinguish between reason and nature. For how would such a theoretical approach even look like? For Habermas, the negation of the essential separation of reason and nature seems to be tantamount to anti-Enlightenment, perhaps even panpsychist views.[17] Hence, he rather adheres to a terminological distinction that may not be absolutely flawless, and thus provides reason with a

---

[16]Cf. Habermas (1995, 514ff.). Ott (2010, 91) points out that Habermas (1991, 226) in his "Remarks on Discourse Ethics" addresses the aesthetic and moral perception of "nature itself" but remains vague about any ethical consequences.

[17]Arguably demonstrated by the fact that Habermas (1995, 505) gives the relevant chapter in *Vorstudien und Ergänzungen zur Theorie des kommunikativen Handelns* (*On the Pragmatics of Social Interaction: Preliminary Studies in the Theory of Communicative Action*, 2000) the title "Reason and nature – a reconciliation at the cost of reenchantment?"

secure sphere of validity, in which one has to struggle for self-determination and fight heteronomy in the name of humanity.

However, the figure of thought that sets "possibly threatened humanity here" against "the threatening forces there" can easily lead to a terminology which accepts that certain dimensions of experience are not represented, or can only be represented when the terms are used rather loosely. This can be seen clearly when Habermas addresses the mind—body problem. Rejecting monist approaches, Habermas draws a fundamental distinction between mind and nature. Though for him the point of this distinction is not ontological but methodological. Hence he has no interest in declaring mind and nature to be distinct substances. Instead, he suggests an understanding of the ways the humanities and the natural sciences approach the world as two distinct and nonreducible *language games*. Those deal, Habermas elaborates, for the humanities, with the communicative structure of human life, and for the natural sciences, with things and events that can be experienced (cf. Habermas 2008, 166ff.). Neither language game can be reduced to the other, revealing an inherently dualistic conception of the world, which Habermas sees purely pragmatic: "The fact that the one language game cannot be reduced to the other need not bother us any more than the fact that one tool cannot be replaced by another" (Habermas 2008, 165). After all, both language games originate from "the comprehensive rationality which", so Habermas in an early work written during the positivism dispute, "in the natural hermeneutics of everyday language, is still, as it were, naturally at work. In the sciences, however, this rationality must be re-established between the now-separated moments of formalized language and objectivized experience by means of critical discussion" (Habermas 1976a, 219f.).

To be able to counter the instrumentalizing forces of objectivized science and technical rationale, Habermas emphasizes that rationality is always based on the language and realized in communication. But communicative action needs to be comprehended, which is only possible from the inside perspective of potential participants of communication, and with hermeneutical methods. For this methodology, Habermas accepts that he may lose sight of the fact that the world quite often calls for our attention in rather nontheoretical ways and without being mediated by language. It is when confronted with the unexpected that we are encouraged or even forced to change our ideas about the world and at best, adjust them closer to reality. This does not to mean that a case is made for a strict ontological separation of a world of facts and a world of norms, as Habermas and other like-minded critical theorists objected (cf. Habermas 1976a, 215–220). But it explains our concerns when in the justified criticism of empiricism, the whole world is treated as only accessible by means of language and theory.[18]

At this point, the debate with Critical Theory cannot be brought to a satisfying end. It would need an alternative proposal, one that takes up the aspects of Critical Theory that correspond with the goals of sustainability discourse, but does not neglect the concerns described especially in response to Habermas' theoretical framework. In conclusion, it cannot be denied that there is a more than passing family resemblance between Critical Theory and sustainability discourse. They both aim to not just be a function of society but to reflect on this fact, and to incorporate this reflection into their own scientific practice. Both develop their critical standards from a regulative idea—the critical theorists from the idea of a free and emancipated life, the sustainability researchers from the idea of a sustainable society.[19] Both aim at establishing normative statements. It does seem as if sustainability discourse could be realized as Critical Theory, if not for the problem that the relationship between nature and society needs to be addressed. This problem is, it seems, mostly relevant to empirical research that provides knowledge for a sustainable development of society. It obviously does not concern the ethical discourse. For a normative theory of sustainability can very well be based on Habermas' discourse ethics (cf. Ott and Döring 2008). It is empirical science which cannot afford to ignore certain perceptual experiences and thus risk to not adequately conceptualize its research object and data sources. If—to satisfy the conceptual scheme—potential experiences do not have the chance to become theoretically relevant,

---

[18]Cf. Habermas (1976a, 203f). Unlike Adorno, who ascribes knowledge processes to the existence of the non-identical, cf. Görg (1999, 129).

[19]This is at least emphasized by sustainability researchers whose idea of sustainable social development does not stop at the availability of resources (Norton 2005; Burger 2006; Martens 2006).

the normative foundation of empirical science is undermined and its contribution to sustainability discourse rendered baseless. It makes no difference that "freedom" and "a humane society" are the goals of such a theory. Habermas' conceptual framework sets an a priori restriction on the possible relations between society and nature. Which is why it cannot be the starting point for research aiming to generate knowledge about society and its embeddedness in the physical world.[20]

To Understand and Overcome Paralysis

Having analyzed what strategies social sciences use to react to its subject's ambiguity, the results can be condensed into a thesis: The sustainability discourse creates a so-called double-bind situation for social sciences (cf. Bateson 2000, 201–278). This double bind prevents them to get seriously involved in sustainability discourse or at least make indicatory contributions to it. What double bind do I mean? There is, on the one hand, the self-image that sees social sciences as a result of the discovery that the social order can be actively shaped and in fact changed. It is this self-image that brought social sciences public recognition. Put more solemnly, social sciences represent the ability of people to put fate in their hand and shape the world. Because of this, and because an independent science needs a constitutive research subject of its own, social scientists are practically forced to defend the autonomy of the social, against attempts to bring nonsocial factors to the discussion, as significant causes which define the shape of the social order.[21]

On the other hand, there is an expectation within sustainability discourse that the social sciences should address the limits to which

---

[20]Similar to Habermas, the older Critical Theory devalues experiences in favor of theory, which is always considered paramount. Hence the opportunity is lost to systematically expedite the reevaluation of one's own standards (cf. Horkheimer 1988, 212f.). Admittedly, there are passages when, for instance, nature is addressed as an independent and recalcitrant factor. But in general, it needs to be stated that both Adorno and Horkheimer share a deep unease about the fact that society is exposed to natural processes (cf. Görg 1999, 114–133, esp. 128f.).

[21]This explains, too, why the allegation that a specific sociological position uses naturalistic, biologistic, or metaphysical arguments—i.e. that it refers to non-social factors—is one the most severe allegations in social science debates.

societies can be shaped and developed—limits set by nature. And therefore those social sciences, which come with predominantly reflexive or even emancipatory intentions, simply do not join in. They pretend that "sustainable development" is not their business, as if their own research about social structures, education issues, family developments, democracy, etc. did not explicitly and implicitly deal with questions about the future of society. Sustainability discourse is seen as a discourse about the scarcity of environmental resources, addressed to specialized subdisciplines such as environmental sociology or environmental law. These subdisciplines have no other choice; they cannot wholly refuse to deal with the subject. Sustainability has, after all, to do with the environment and nature, and this is their area of expertise. But their standing within the social sciences would come under attack if they started to develop conceptual frameworks, explaining the state and the development of social orders not only with social factors, but also with so-called natural factors. This paradox results in a situation where indifferent social scientists sit by and watch idly, while other disciplines discuss sustainability.

How can we overcome this paralysis? Definitely not by suggesting social scientists no longer employ social phenomena to explain other social phenomena (like e.g. Burger 2007). Such a suggestion obviously aims to clarify that the forms of human coexistence in the concrete world are always also determined by natural conditions and that they have a physical dimension. But it completely disregards the prevalent self-image of social scientists. As discussed above, this self-image is soundly based on the idea that social sciences have their own subject, which they tackle with their own scientific terminology and methodology.

However, even though the autonomy of the social can be experienced, it is yet to be decided whether the social, following a constitutional logic, needs to be understood as autonomous and hence, that its respective form is determined by nothing but its own past. Alternatively, talking about the "autonomy of the social" may also be understood as a metaphor. The metaphorical constitutes the transfer of an ethical category—free will—into the sphere of the philosophy and methodology of science, a transfer encouraged by political philosophy and by social groups struggling for sovereignty.

What then is this category transferred to? Let me use an illustration from everyday language. In the sentence, "In Switzerland, trains are well organized and punctual," the word "Switzerland" refers to a country, meaning a conglomerate of complex sets of social rules and distinct natural conditions: a space that can be described geographically, legally defined as the territory of a state, internationally recognized, where people live who call themselves "Swiss", with mountainous, inhospitable natural conditions, internally defined as a confederacy of mostly independent political communities with their own regional territories, with major centers of trade and—most importantly—of commerce. These and other characteristics are the ones we associate with the word "Switzerland". But when we (half enviously, half admiringly in Germany) talk about the railway system in Switzerland and emphasize how well its transport structure accommodates passengers' needs, we focus solely on the sets of social rules. We may even know in the back of our minds that the natural conditions in Switzerland pose specific challenges for building and maintaining a transport infrastructure. But in actuality, we disregard the physical world and refer to institutions. In terms of methodology, we talk about Switzerland as an abstraction, addressing an interdependency of social causes and social effects that can be experienced autonomously.

But what if the subject of social sciences—"the social", or "the society"—was only methodologically conceptualized as an autonomous factor? Then, the question of whether the social, or society, is autonomous or not, could become a research question of an interdisciplinary discourse. There would be no grounds for a prior ontologization of the social. But this would not put social scientists onto the defensive, quite the contrary. For the crucial question then is no longer *if* but *to what extent* the concrete world, its forms and developments, can be explained with a specific interdependence of functions and contexts. Not the acknowledgment of a belief stands at the beginning of a social science study, but the discussion of perceptual content that is considered to be "socially contingent".

This does not, however, mean that social scientists need no longer be familiar with important doctrines and philosophical and ontological controversies in the social sciences. Following Emile Durkheim's dictum, they should not choose "between the great hypotheses that divide metaphysicians" (Durkheim 1995, 218; translation J.J.). But they should know the hypotheses. Otherwise metaphysics will catch up with them behind their backs, like it inadvertently does with all empirical scientists. The conceptual apparatuses we use to interpret experience and turn it into knowledge, presupposes images of the world, of humanity, and of nature, all of which take a stand on what constitutes humanity—whether and how one can differentiate in a meaningful way between mind and nature, or mind and matter; whether there is free will, and so on. The answers to such questions can be found in the views that inform the conceptual frameworks of science. And of course, this is true for other conceptual frameworks as well. The terms we use in everyday life are just as much embedded in religious and ideological ideas. And in scientific discourse as much as in everyday life, answers may become flawed and no longer convincing. In view of the relationship between mind and nature, it is not only man-made environmental problems that challenge us to search for new answers. The genderedness of human life in particular and human corporeality, in general, have become subjects of public controversies—not last owed to new diagnostic and therapeutical possibilities. This again has led to critical inquiries of scientific concepts and their teleological implications (cf. Haraway 1985; Merchant 1980; Scheich 1993; Habermas 2001; Agamben 1998). To be able turn such inquiries into research questions, social sciences need a conceptual framework focussing on cultural, psychic, and social factors as much as on human relationships with physicochemical and biological systems. And it needs a methodology that enables them to research and investigate these relationships. Thus equipped, I argue, knowledge can be produced in the context of a coevolutionary science, which will provide sustainability discourse with adequate decision-making tools. I discuss this in more detail in the next chapter.

Toward an Interdisciplinary Coevolutionary Science: An Outline

In what follows, "coevolutionary science" is defined *not* as a clearly distinguished science discipline with a concrete research subject, but as a scientific research program. The term combines heterogeneous research activities within and across various disciplines, wherever scientists examine changes of their specific research subjects, and ask whether these changes occurred because of causes lying within the subjects or because of outside causes, and whether these changes have consequences for other social, cultural, and physical contexts. Thus, there already exists a coevolutionary practice, in so far as the relationship between autonomy and reciprocity is examined in the development of one or several systems or spheres. Such studies exist e.g. for several biological species (cf. Ehrlich and Raven 1964), for physicochemical and biological systems such as climate and organisms (cf. Schneider and Londer 1984), for various systems of action, e.g. of society and its state (cf. Böhret and Konzendorf 1997), of science, technology, and society (cf. Rip 2002), of religious motivational structures and economic systems (cf. Weber 2002), or even for action systems with physicochemical and biological systems like the exploitation and use of fossil energy sources by modern societies and the climate (cf. Flannery 2005). Because of the abstract terminology, a wide and barely manageable range of cases is covered, and in today's scientific landscape those studying the interdependence between action systems and physicochemical and biological systems certainly carry the biggest potential for controversy.

For the coevolutionary point of view, it is not important which spheres or systems are considered interdependent entities that dynamize each other. Rather, it is of importance that the relationship between the entities results in the continuous emergence of new structures. Terms like "couplings", "interfaces", or "network of relations" leave such interdependencies under-determined. These terms only describe what everyone says—that somehow everything is related. But the interesting question for a scientist is: *How* is it related? For studies aiming at answering this question, the concept of coevolution formulates an epistemic goal.

This should, however, not be taken to mean that coevolutions are viewed as a desirable outcome, as progress or as a step in the right direction, or are regarded as something good in a moral sense. "Epistemic goal" rather means that "coevolution" is a productive criterion for scientific research, as the establishment of a coevolutionary relationship between two different systems or spheres poses the ultimate challenge for the empirical sciences. To put it differently: Looking for coevolutions means to focus on linear as well as nonlinear relationships, on independent developments, self-intensifying processes, positive or absent feedback effects, and interdependencies.

The knowledge of a coevolutionary science is of special interest for sustainability discourse. And not only because it can be a safeguard against one-dimensional recipes for social organization, against planning optimism and faith in progress (cf. Norgaard 1994). Of course, this knowledge can, in turn, be used for the shaping of the world and societies, and thus give a higher validity to projections, scenarios, and predictions of the future. In this sense, coevolutionary science serves as the function in a society that is able to learn by engaging with its environment and its own circumstances (cf. Eder 1999; Habermas 1976b; Siebenhüner and Heinrichs 2010; Berkes et al. 2003)—hopefully through thought experiments and research-based discussions first, before societies start social experiments in reality (cf. Groß et al. 2005), at the risk of determining structures for a long time and perhaps irreversibly.

If research for sustainable development is to take into account possible coevolutions, a conceptual framework is needed that focuses on action and society *as autonomous research subjects* and, at the same time, allows researchers to study them *within their complex interrelations*. Like a map, this conceptual framework should objectively depict the inner structure of social and action systems as well as formally present possible external relations. This gives researchers opportunity to reconsider various positions about questions of sustainable development, including their own. Here I follow Bourdieu's ideal of an objective science that at the same time makes visible the researcher's position (cf. Bourdieu 1997).

The concepts and philosophical references of the conceptual framework itself are by no means arbitrary. Yet, different philosophical and sociological traditions may be used for such a project.[22] From a methodological point of view, it is only important to say goodbye to the notion that unquestionable knowledge about the experienceable, refractory world could really exist.

To say goodbye to this notion does not mean, however, that from now on all criteria for the truth or adequacy of a statement are no longer valid. It means to create knowledge and insights that are "reliable" according to the prevailing, generally accepted criteria of rationality. This may seem but a small shift of emphasis but it is of immense importance. For this shift urges researchers to take criteria of rationality into account when they evaluate knowledge. Statements resulting from formal mathematically models are no longer automatically considered unquestionable knowledge. And knowledge about society can be seen as "reliable" as well, claiming the same degree of reliability as knowledge about nature. More generally, the idea that it is possible to distinguish between the search for knowledge on the one hand, and creative action on the other, is undermined. For, as pointed out above, all action refers to knowledge when a situation gets defined. Moreover, new knowledge is produced when actions are reflected upon.

It goes hand in hand with this shift in the evaluation that I don't apostrophize my conceptual clarification for coevolutionary research as a "theory of nature-society relations". I rather see it as a preliminary work for a methodology which helps researchers in the field of sustainability studies to assess the adequacy of research questions, applied methods, etc. I hope a fully developed coevolutionary science will produce empirical knowledge that can be used as an instrument for scientific

---

[22]My version is based on the classical pragmatism and, as its concept of action is not sufficiently sophisticated, on functionalist action theory. This combination may seem unusual but both approaches complement each other very well, if the goal is to develop a concept of social sciences that takes into account the uniqueness of their research subject and can rival the scientific character of the "hard" natural sciences (cf. Jetzkowitz 2003).

research, and will reflect to what extent this knowledge may serve the goal of moving present society toward sustainability.[23]

If sustainability discourse is equipped with a coevolutionary conceptual framework and a methodology, then an interrelation of sustainability research and sustainability discourse becomes possible. On the one hand, research helps discourse to assess the future viability of potential social changes, and, building on this, to prepare binding decisions about standards to be applied and goals to be politically aspired. On the other hand, participants in sustainability discourse would be able to evaluate, critically and on the basis of formal criteria, studies for a sustainable development, and to encourage the search of new knowledge by pointing out concrete aims and strategies.[24]

Much would be gained if a conceptual framework and its related methodology could thus indeed feed questions and problem definitions of the general public back to the professionalized sciences.[25] Knowledge production as an essential factor in the shaping of societies could no longer be legitimized away from the public eye. Questions about the meaningfulness and relevance of scientific inquiry would be given greater weight. And the idea that a society can control its own development through knowledge would be seen as more realistic. For such a conceptual framework and its related methodology of a, mind you, interdisciplinary research program necessarily addresses a lay audience which may be scientifically educated but should not need specialized disciplinary knowledge to be able to follow an argument.[26]

---

[23]Here I address Horkheimer's (1988) request that criteria of reason guide the goals of scientific research as much as the scientific process.

[24]Burger and Christen (2011) have a similar concern when they distil from sustainability discourse adequacy conditions for sustainability conceptions.

[25]Efforts towards a transdisciplinary science that puts a stronger emphasis on offering solutions for social problems (rather than homemade ones resulting from discourses of the science disciplines) also are provided with an orienting concept to assess cooperations between science and social practice.

[26]In this context connections between coevolutionary science to concepts of deliberative politics (cf. e.g. Habermas 2015) should be discussed.

In the following chapter, I discuss how coevolutionary science positions itself to common views of knowledge about nature and society.[27] Such elementary considerations are commonly found in classic works of the social sciences but they have mostly gone out of fashion. This makes sense, insofar as discourses of philosophy—including discourses of the philosophy of science have continuously become more sophisticated and specialized. Most people from outside of the field can no longer comprehend them, much less evaluate them with due diligence. All too easily one may get lost in the depths of discourse without apparent gain for own research. I still enter the philosophical territory, not as a reminiscence of the classics, but because of the unclear state of current knowledge. For this has been one of the points of my deliberations above: The conceptualization of the research subject—as interfaces of or interrelations between nature and society—as well as the diverging knowledge traditions of the natural sciences and the social sciences constitute an essential part of the problematic of a sustainable development of society. It is for this reason that hereinafter I focus on the relationship between knowledge and action, and discuss how the kind of coevolutionary science sketched above may solve the knowledge problems of sustainability discourse.

References

Agamben, G. (1998 [1995]). *Homo Sacer: Sovereign Power and Bare Life* (D. Heller-Roazen, Trans.). Stanford: Stanford University Press.

Bateson, G. (2000 [1972]). *Steps to an Ecology of Mind: Collected Essays in Anthropology, Psychiatry, Evolution, and Epistemology*. Chicago, IL: University of Chicago Press.

---

[27]Gesa Lindemann (2009, 1; translation J.J.) defines this dimension of sociological theorizing as "social theory": "The term social theory refers to those aspects of sociological theory that determine what is to be understood as a social phenomenon and what methodological principles are to be used in the collection and analysis of data." I suggest a different definition of sociality and use a different terminology. But I agree with Lindemann that all theories in social sciences (and in fact all scientific theories) necessarily include assumptions about their research subject and how to approach it.

Berkes, F., Colding, J., & Folke, C. (Eds.). (2003). *Navigating Social-Ecological Systems. Building Resilience for Complexity and Change.* Cambridge: Cambridge University Press.

Bettencourt, L. M. A., & Kaur, J. (2011). Evolution and Structure of Sustainability Science. *PNAS, 108*(49), 19540–19545.

Böhret, C., & Konzendorf, G. (1997). *Ko-Evolution von Gesellschaft und funktionalem Staat. Ein Beitrag zur Theorie der Politik.* Opladen: Westdeutscher Verlag.

Boole, G. (1854). *An Investigation of the Laws of Thought, on Which Are Founded the Mathematical Theories of Logic and Probabilities.* New York: Dover.

Bourdieu, P. (Ed.). (1997). *The Weight of the World: Social Suffering in Contemporary Society.* Cambridge: Polity.

Brulle, R. J. (2000). *Agency, Democracy, and Nature: The U.S. Environmental Movement Organizations from a Critical Theory Perspective.* Cambridge, MA: MIT Press.

Burger, P. (2006). *Sustainability Science: The Science of the Future.* Unpublished Manuscript.

Burger, P. (2007). Nachhaltigkeitstheorie als Gesellschaftstheorie. Ein philosophisches Plädoyer. In Schweizerische Akademie der Geistes- und Sozialwissenschaften (Ed.), *Nachhaltigkeitsforschung – Perspektiven der Sozial- und Geisteswissenschaften* (pp. 13–34). Bern: Schweizerische Akademie der Geistes- und Sozialwissenschaften.

Burger, P., & Christen, M. (2011). Towards a Capability Approach of Sustainability. *Journal of Cleaner Production, 19*(8), 787–795.

Carson, R. (1962). *Silent Spring.* Greenwich, CT: Fawcett.

Commoner, B. (1971). *The Closing Circle: Nature, Man, and Technology.* New York: Knopf.

Costanza, R., Cumberland, J., Daly, H., Goodland, R., & Norgaard, R. (1997). *An Introduction to Ecological Economics.* Boca Raton, FL: CRC Press LLC.

Daly, H. E. (1973). The Steady State Economy: Toward a Political Economy of Biophysical Equilibrium and Moral Growth. In H. E. Daly (Ed.), *Toward a Steady State Economy* (pp. 149–174). San Francisco: W.H. Freeman.

Daly, H. E. (1987). The Economic Growth Debate: What Some Economists Have Learned but Many Have Not. *Journal of Environmental Economics and Management, 14,* 323–336.

Daly, H. E. (2008). *A Steady-State Economy.* Report to the Sustainable Development Commission, UK (April 24, 2008). Retrieved November

16, 2013, from http://www.sd-commission.org.uk/data/files/publications/ Herman_Daly_thinkpiece.pdf.

Dickens, P. (2004). *Society and Nature: Changing Our Environment, Changing Ourselves*. Cambridge: Polity Press.

Didham, R. J., & Ofei-Manu, P. (2015). Social Learning for Sustainability. In V. W. Thoresen, D. Doyle, J. Klein, & R. J. Didham (Eds.), *Responsible Living: Concepts, Education and Future Perspectives* (pp. 233–252). Cham: Springer.

Dryzek, J. S. (1987). *Rational Ecology, Environment and Political Economy*. Oxford: Basil Blackwell.

Durkheim, E. (1995 [1895]). *Die Regeln der soziologischen Methode* (R. König, Ed. and Intro.). Frankfurt am Main: Suhrkamp.

Dyball, R., Brown, V. A., & Keen, M. (2007). Towards Sustainability: Five Strands of Social Learning. In A. E. J. Wals (Ed.), *Social Learning: Towards a Sustainable World* (pp. 181–194). Wageningen: Wageningen Academic Publishers.

Eckersley, R. (1992). *Environmentalism and Political Theory: Toward an Ecocentric Approach*. Albany, NY: Suny Press.

Eckersley, R. (1994). Wo bleibt die Emanzipation der Natur? Habermas' kritische Theorie aus ökozentrischer Sicht. In W. Zierhofer & D. Steiner (Eds.), *Vernunft angesichts der Umweltzerstörung* (pp. 119–158). Opladen: Westdeutscher Verlag.

Eder, K. (1999). Societies Learn and Yet the World Is Hard to Change. *European Journal of Social Theory, 2*, 195–215.

Ehrlich, P. R., & Raven, P. H. (1964). Butterflies and Plants: A Study in Coevolution. *Evolution, 18*, 586–608.

Engelhardt, A., & Kajetzke, L. (Eds.). (2010). *Handbuch Wissensgesellschaft. Theorien, Themen und Probleme*. Bielefeld: transcript.

Engels, F. (1979 [1892]). Engels an Laura Lafargue in Le Perreux. In K. Marx & F. Engels, *Werke* (Band 38, pp. 544–546). Berlin: Dietz-Verlag.

Flannery, T. (2005). *The Weather Makers: The History and Future Impact of Climate Change*. London: Allen Lane.

Gay, P. (1967/70). *The Enlightenment: An Interpretation*. London: Weidenfeld & Nicolson.

Geels, F. W., & Schot, J. (2007). Typology of Sociotechnical Transition Pathways. *Research Policy, 36*(3), 399–417.

Görg, C. (1999). *Gesellschaftliche Naturverhältnisse*. Münster: Westfälisches Dampfboot.

Grober, U. (2010). *Die Entdeckung der Nachhaltigkeit. Kulturgeschichte eines Begriffs.* München: Verlag Antje Kunstmann [also published in English: Grober, U. (2012). *Sustainability: A Cultural History* (R. Cunningham, Trans.). Totnes, Devon: Green Books].

Groß, M. (2001). *Die Natur der Gesellschaft. Eine Geschichte der Umweltsoziologie.* Juventa: Weinheim, München.

Groß, M., Hoffmann-Riem, H., & Krohn, W. (2005). *Realexperimente. Ökologische Gestaltungsprozesse in der Wissensgesellschaft.* Bielefeld: transcript.

Habermas, J. (1976a [1964]). A Positivistically Bisected Rationalism. In T. W. Adorno, H. Albert, R. Dahrendorf, J. Haber, H. Pilot, & K. R. Popper (Eds.), *The Positivist Dispute in German Sociology* (G. Adey & D. Frisby, Trans.) (pp. 198–225). London: Heinemann.

Habermas, J. (1976b). *Zur Rekonstruktion des Historischen Materialismus.* Frankfurt am Main: Suhrkamp.

Habermas, J. (1981a). *Theorie des kommunikativen Handelns. Band 1: Handlungsrationalität und gesellschaftliche Rationalisierung.* Frankfurt am Main: Suhrkamp [also published in English: Habermas, J. (1984). *Theory of Communicative Action, Volume One: Reason and the Rationalization of Society* (T. A. McCarthy, Trans.). Boston, MA: Beacon Press].

Habermas, J. (1981b). *Theorie des kommunikativen Handelns. Band 2: Zur Kritik der funktionalistischen Vernunft.* Frankfurt am Main: Suhrkamp [also published in English: Habermas, J. (1987). *Theory of Communicative Action, Volume Two: Lifeworld and System: A Critique of Functionalist Reason* (T. A. McCarthy, Trans.). Boston, MA: Beacon Press].

Habermas, Jürgen. (1991). *Erläuterungen zur Diskursethik.* Frankfurt am Main: Suhrkamp.

Habermas, J. (1995). *Vorstudien und Ergänzungen zur Theorie des kommunikativen Handelns.* Frankfurt am Main: Suhrkamp.

Habermas, J. (2001). *Die Zukunft der menschlichen Natur. Auf dem Weg zur liberalen Eugenik?* Frankfurt am Main: Suhrkamp [also published in English: Habermas, J. (2003). *The Future of Human Nature.* Cambridge: Polity Press].

Habermas, J. (2008). *Between Naturalism and Religion: Philosophical Essays.* Cambridge, UK and Malden, MA: Polity Press.

Habermas, J. (2015 [1994]). *Between Facts and Norms: Contributions to a Discourse Theory of Law and Democracy.* New York, NY: Wiley.

Haraway, D. (1985). A Manifesto for Cyborgs: Science, Technology, and Socialist Feminism in the 1980s. *Socialist Review, 80,* 65–108.

Heidenreich, M. (2003). Die Debatte um die Wissensgesellschaft. In S. Böschen & I. Schulz-Schaeffner (Eds.), *Wissenschaft in der Wissensgesellschaft* (pp. 25–51). Opladen: Westdeutscher Verlag.

Horkheimer, M. (1988 [1937]). Traditionelle und kritische Theorie. In M. Horkheimer & G. Schriften, *Band 4: Schriften 1936–1941* (pp. 162–216). Frankfurt am Main: Suhrkamp.

Horkheimer, M., & Adorno, T. W. (1969 [1944]). *Dialektik der Aufklärung. Philosophische Fragmente*. Frankfurt am Main: Fischer.

Huber, J. (1995). *Nachhaltige Entwicklung*. Edition Sigma: Strategien für eine ökologische und soziale Erdpolitik. Berlin.

Huber, J. (2000). Industrielle Ökologie. Konsistenz, Effizienz und Suffizienz in zyklusanalytischer Betrachtung. In R. Kreibich & U. E. Simonis (Eds.), *Global Change* (pp. 109–126). Berlin: Verlag Arno Spitz.

Huber, J. (2011). Ökologische Modernisierung und Umweltinnovation. In M. Groß (Ed.), *Handbuch Umweltsoziologie* (pp. 279–302). Wiesbaden: VS Verlag.

Jetzkowitz, J. (2003). Funktionale Analyse als Zeichenprozess – Parsons' Soziologie als Theorie semiotischer Subjekte. In J. Jetzkowitz & C. Stark (Eds.), *Soziologischer Funktionalismus. Zur Methodologie einer Theorietradition* (141–176). Opladen: Leske+Budrich.

Jetzkowitz, J. (2008). Die Anpassung an den Klimawandel im Blick der Gesellschaftstheorie. In Institut WAR (Ed.), *Klimawandel – Markt für Strategien und Technologien?!* (pp. 99–114). Darmstadt: Schriftenreihe WAR Nr.196.

Jüdes, U. (1997). Nachhaltige Sprachverwirrung. Auf der Suche nach einer Theorie des Sustainable Development. *Politische Ökologie, 15*(52), 26–29.

Kant, I. (1899 [1781/1787]). *The Critique of Pure Reason* (J. M. D. Meiklejohn, Trans.). New York, NY: Wiley.

Kates, R. W., Clark, W. C., Corell, R., Hall, J. M., Jaeger, C. C., Lowe, I., et al. (2001). Sustainability Science. *Science, 292*, 641–642.

Krämer, W., & Mackenthun, G. (2001). *Die Panik-Macher*. München: Piper.

Kroll, G. (2001). The "Silent Springs" of Rachel Carson: Mass Media and the Origins of Modern Environmentalism. *Public Understanding of Science, 10*(4), 403–420.

Lear, L. (1997). *Rachel Carson. Witness for Nature*. London: Penguin Books.

Liedman, S.-E. (1998). Engels and the Laws of Dialectics. In R. Panasiuk & L. Nowak (Eds.), *Marx's Theories Today* (15–35). Amsterdam: Rodopi B.V.

Lindemann, G. (2009). *Das Soziale von seinen Grenzen her denken*. Weilerswist: Velbrück.

Linse, U. (1986). *Ökopax und Anarchie. Eine Geschichte der ökologischen Bewegungen in Deutschland*. München: dtv.

Lomborg, B. (2001). *The Skeptical Environmentalist*. Cambridge and New York: Cambridge University Press.

Lucas, E. (1964). Marx' und Engels' Auseinandersetzung mit Darwin: zur Differenz zwischen Marx und Engels. *International Review of Social History, 9*, 433–469.

Luederitz, C., Schäpke, N., Wiek, A., Lang, D. J., Bergmann, M., Bos, J. J., et al. 2016. Learning Through Evaluation: A Tentative Evaluative Scheme for Sustainability Transition Experiments. *Journal of Cleaner Production*, http://dx.doi.org/10.1016/j.jclepro.2016.09.005.

Luke, T. W., & White, S. K. (1985). Critical Theory, the Informational Revolution, and an Ecological Path to Modernity. In J. Forester (Ed.), *Critical Theory and Public Life* (pp. 22–53). Cambridge, MA: MIT Press.

Lutzenhiser, L. (1993). Social and Behavioral Aspects of Energy Use. *Annual Review of Energy and the Environment, 18*, 247–289.

Martens, P. (2006). Sustainability: Science or Fiction? *Sustainability: Science, Practice and Policy, 2*(1), 1–5.

Marx, K. (1887). *Capital. A Critique of Political Economy, Volume I, Book One: The Process of Production of Capital*. Moscow: Progress Publishers.

Marx, K., & Engels, F. (1977 [1888]). *Manifesto of the Communist Party*. Moscow: Progress Publishers.

Marx, K., & Engels, F. (1978 [1850/1885]). Ansprache der Zentralbehörde an den Bund vom März 1850. In K. Marx & F. Engels, *Werke* (Band 7, pp. 244–254). Berlin: Dietz-Verlag.

Maxeiner, D., & Miersch, M. (1999). *Lexikon der Öko-Irrtümer*. Frankfurt am Main: Eichborn.

Meadows, D. (2000). Es ist zu spät für eine nachhaltige Entwicklung. Nun müssen wir für eine das Überleben sichernde Entwicklung kämpfen. In W. Krull (Ed.), *Zukunftsstreit* (pp. 125–149). Weilerswist: Velbrück.

Meadows, D. L., Meadows, D. H., Randers, J., & Behrens III, W. W. (1972). *The Limits to Growth: A Report for the Club of Rome's Project on the Predicament of Mankind*. New York: Universe Books.

Merchant, C. (1980). *The Death of Nature: Women, Ecology, and the Scientific Revolution*. San Francisco: Harper & Row.

Murcott, S. (1997). Sustainable Development: A Meta-Review of Definitions, Principles, Criteria, Indicators, Conceptual Frameworks, Information Systems. In *Annual Conference of the American Association for the Advancement of Science. IIASA Symposium on "Sustainability Indicators".* Seattle, WA, February 13–18, 1997.

Naess, A. (1973). The Shallow and the Deep, Long-Range Ecology Movement: A Summary. *Inquiry, 16*(1), 95–100.

Nerlich, B. (2003). Tracking the Fate of the Metaphor *Silent Spring* in British Environmental Discourse. Towards an Evolutionary Ecology of Metaphor. In *Metaphoric* 04/2003, 115–140.

Norgaard, R. B. (1994). *Development Betrayed: The End of Progress and a Coevolutionary Revisioning of the Future.* London: Routledge.

Norton, B. G. (2005). *Sustainability. A Philosophy of Adaptive Ecosystem Management.* Chicago: The University of Chicago Press.

Nowotny, H. (2001). Vom Geschichtenerzählen zur Koevolutionswissenschaft. *GAIA, 10*(4), 262–264.

Ott, K. (1993). *Ökologie und Ethik: Ein Versuch praktischer Philosophie.* Tübingen: Attempto-Verlag.

Ott, K. (2010). *Umweltethik zur Einführung.* Marburg: Metropolis-Verlag.

Ott, K., & Döring, R. (2008). *Theorie und Praxis starker Nachhaltigkeit.* Marburg: Metropolis-Verlag.

Outram, D. (1995). *The Enlightenment.* Cambridge: Cambridge University Press.

Porter, R. (2001). *The Enlightenment.* Basingstoke: Palgrave.

Radkau, J. (2008). *Nature and Power: A Global History of the Environment* (T. Dunlap, Trans.). Cambridge: Cambridge University Press.

Radkau, J. (2011). *Die Ära der Ökologie. Eine Weltgeschichte.* München: Beck [also published in English: Radkau, J. (2014). *The Age of Ecology: A Global History.* Cambridge: Polity].

Radnitzky, G. (1989). Wissenschaftstheorie, Methodologie. In H. Seiffert & G. Radnitzky (Eds.), *Handlexikon zur Wissenschaftstheorie* (pp. 463–471). München: Ehrenwirth.

Reichholf, J. H. (2002). *Die falschen Propheten. Unsere Lust an Katastrophen.* Berlin: Klaus Wagenbach Verlag.

Renn, O. (1985). Die alternative Bewegung: Eine historisch-soziologische Analyse des Protestes gegen die Industriegesellschaft. *Zeitschrift für Politik, 32,* 152–194.

Rip, A. (2002). Co-Evolution of Science, Technology and Society. An Expert Review for the Bundesministerium Bildung und Forschung's Förderinitiative Politik, Wissenschaft und Gesellschaft (Science Policy Studies), as managed by the Berlin-Brandenburgische Akademie der Wissenschaften. Enschede: University of Twente, June 2002. Retrieved March 19, 2014, from http://citeseerx.ist.psu.edu/viewdoc/download?doi=10.1.1.201.6112&rep=rep1&type=pdf.

Røpke, I. (2004). The Early History of Modern Ecological Economics. *Ecological Economics, 50*(3/4), 293–314.

Røpke, I. (2005). Trends in the Development of Ecological Economics from the Late 1980s to the Early 2000s. *Ecological Economics, 55*(2), 262–290.

Scheich, E. (1993). *Naturbeherrschung und Weiblichkeit. Denkformen und Phantasmen der modernen Naturwissenschaften.* Pfaffenweiler: Centaurus.

Schellnhuber, H.-J. (2001). Die Koevolution von Natur, Gesellschaft und Wissenschaft – Eine Dreiecksbeziehung wird kritisch. *GAIA, 10*(4), 258–262.

Scheunemann, E. (2008). *Vom Denken der Natur. Natur und Gesellschaft bei Habermas.* Norderstedt: Books on Demand.

Schneider, S. H., & Londer, R. (1984). *Coevolution of Climate and Life.* San Francisco: Sierra Club Books.

Sen, A. (1987). *On Ethics and Economics.* Oxford: Basil Blackwell.

Siebenhüner, B., & Heinrichs, H. (2010). Knowledge and Social Learning for Sustainable Development. In M. Gross & H. Heinrichs (Eds.), *Environmental Sociology: European Perspectives and Interdisciplinary Challenges* (pp. 185–199). Doderecht: Springer.

Sieferle, R. P. (1984). *Fortschrittsfeinde? Opposition gegen Technik und Industrie von der Romantik bis zur Gegenwart.* München: Beck.

Sitter-Liver, B. (2000). Tiefen-Ökologie: Kontrapunkt im aktuellen Kulturgeschehen. *Natur und Kultur, 1*(1), 70–88.

Stehr, N. (1994). *Knowledge Societies.* London: Sage.

Theobald, W. (2003). *Mythos Natur. Die geistigen Grundlagen der Umweltbewegung.* Wissenschaftliche Buchgesellschaft: Darmstadt.

Thomas, W. I., & Thomas, D. S. (1928). *The Child in America: Behavior Problems and Programs.* New York: Knopf.

Ulrich, P. (1986). *Transformationen der ökonomischen Vernunft. Fortschrittsperspektiven der modernen Industriegesellschaft.* Bern: Paul Haupt.

Ulrich, P. (2008). Integrative Wirtschaftsethik. Grundlagen einer lebensdienlichen Ökonomie. 4., vollständig neu bearb. Aufl. Bern: Haupt [also published in English: Ulrich, P. (2008). *Integrative Economic Ethics: Foundations of a Civilized Market Economy*. Cambridge: Cambridge University Press].

United Nations Educational, Scientific and Cultural Organization (UNESCO). (2005). Towards Knowledge Societies (UNESCO World Report). Paris: UNESCO Publishing, http://unesdoc.unesco.org/images/0014/001418/141843e.pdf.

Vahrenholt, F., & Lüning, S. (2012). *Die kalte Sonne. Warum die Klimakatastrophe nicht stattfindet*. Hoffmann und Campe: Hamburg.

Voß, J.-P., Newig, J., Kastens, B., Monstadt, J., & Nöltig, B. (2007). Steering for Sustainable Development: A Typology of Problems and Strategies with Respect to Ambivalence, Uncertainty and Distributive Power. *Journal of Environmental Policy & Planning, 9*(3/4), 193–212.

von Carlowitz, H. C. (2000 [1713]). *Sylvicultura oeconomica oder haußwirthschaftliche Nachricht und naturmäßige Anweisung zur wilden Baum-Zucht*. Freiberg: TU Bergakademie Freiberg.

von Weizsäcker, E. U. (1994). Erdpolitik. Ökologische Realpolitik an der Schwelle zum Jahrhundert der Umwelt. 4., aktualisierte Aufl., Darmstadt: Wissenschaftliche Buchgesellschaft [also published in English: von Weizsäcker, E. U. (1994). *Earth Politics*. London and Atlantic Highlands, NJ: Zed Books].

Weber, M. (2002 [1905]). *The Protestant Ethic and the Spirit of Capitalism* (S. Kahlberg, Intr. and Trans.). Oxford: Blackwell.

White, L. T. (1967). The Historical Roots of Our Ecologic Crisis. *Science, 155*(3767), 1203–1207.

Wilhite, H. L., & Nørgaard, J. (2004). Equating Efficiency with Reduction: A Self Deception in Energy Policy. *Energy and Environment, 15*(6), 991–1009.

Windelband, W. (1904). *Geschichte und Naturwissenschaft* (3rd ed.). Straßburg: Heitz.

World Commission on Environment and Development (WCED). (1987). Report of the World Commission on Environment and Development: Our Common Future, Chapter 2: Towards Sustainable Development. Retrieved August 26, 2010, from http://www.un-documents.net/ocf-02.htm.

Zierhofer, W. (1994). Ist die kommunikative Vernunft der ökologischen Krise gewachsen? Ein Evaluationsversuch. In W. Zierhofer & D. Steiner (Eds.), *Vernunft angesichts der Umweltzerstörung* (pp. 161–194). Opladen: Westdeutscher Verlag.

2

Where Do We Stand?

Nature, Society and Knowledge—Problems of a Ménage à Trois

How do we know whether climate change is indeed caused by humans? And how is society supposed to respond to the news that man-made environmental changes will lead to the extinction of certain plant or animal species, such as the Siberian tiger? Such questions confront us with problems of knowledge. For we are usually quite careful to keep two fields of knowledge neatly apart. As a matter of course we assume that the natural sciences are different from the social sciences and the humanities. The natural sciences deal with nature, the social sciences with everything produced by human societies—language, history, art, religion, the law. That is, they deal with mind and culture in the broadest sense. The two questions above do not fit into this division of the world, and thus they create problems.

In this chapter, I pursue problems of knowledge that are essential for the discourse of sustainability. I examine why the division of the world in nature and mind (or rather in subjects of the natural sciences and subjects of the social sciences and humanities) makes for problems in

© The Author(s) 2019
J. Jetzkowitz, *Co-Evolution of Nature and Society*,
https://doi.org/10.1007/978-3-319-96652-6_2

the relationship of nature, society, and knowledge. And I ask whether these problems must necessarily arise or if they can somehow be avoided.

Like already mentioned above, I use the philosophy of Immanuel Kant as an illustrative example of how the world appears within the context of a worldview that categorically differentiates nature and mind. Some prior explanations seem necessary. For it was not Kant who invented the separation of nature and mind; he did not even prominently elaborate on it in his work. The philosopher from Königsberg follows a long tradition reaching all the way back to ancient Greece. This tradition is characterized by its definition that what makes humanity human is its differentiation from nature. The *humanum* is "not-nature" (cf. Böhme 1997; Rantis 2004). Both the concept of nature and what constitutes "not-nature" changes over the course of the centuries. Kant, one could argue, is only one among many who used these terms to create their view of the world. One could argue furthermore that Kant did not even mark a specific cesura in this tradition. One look into the history of philosophy provides ample proof of this.

Usually, the European tradition of the nature–mind dualism is divided into three epochs (cf. Gloy 1995, passim, esp. 165f.). In ancient Greece, so the argument, nature is conceived of as predetermined and independent from humans—something that exists outside of the influence of human knowledge and action. The Middle Ages developed, under the influence of Christianity, a concept of nature as creation, from which humanity differs because it was created in the image and the likeness of God. Finally, the modern process of secularization focuses on the humanity's powers of cognition and creation. As producers and constructors, humans are mostly emancipated from nature. Philosophical historiography emphasizes that this new concept of nature and mind (or body and mind) was first established by René Descartes; he is considered to be the first philosopher to break with medieval thought. His philosophy introduces the distinction between *res cogitans* and *res extensa*, that is, the realm of thought that is opposed to the realm of "extension", of not-thought, of matter (cf. e.g. Esfeld 2002, 13–16). The implicit transfer of the mind–nature dichotomy to the epistemological problem is generally seen as a manifestation of the modern worldview. The philosopher no longer primarily raises questions about God and reality, but about the human capacity for knowledge. It becomes the central criterion to distinguish humans from all else. And, building on the separation of the subject and the object of knowledge, an attempt is made to answer the question "What is knowledge".[1]

In the following, I do not talk about Descartes, but not out of a disregard for the assessment of his achievements in philosophical historiography. I chose to talk about Kant because an analysis of his philosophy gives me the chance to discuss basic positions that are important for social sciences in sustainability discourse. Kant's contrary position to the view (much cherished in social sciences, especially in sociology) that we need to be skeptical of the human capacity for knowledge, evolved from ideas inspired by his study of David Hume's empiricist epistemology (cf. Kant 2004, 9ff./A13ff.). Kant's struggle for unquestionable (or certain) knowledge illustrates the dramatic tensions within the interrelated web of nature, society, and knowledge, tensions that also govern sustainability discourse.

Kant, Hume and the Struggle for Certain Knowledge

Let's start again with the same questions as before: How do we know whether climate change is indeed caused by modern civilization? And how is society supposed to respond to the news that the Siberian tiger will soon go extinct? Both questions address the relationship between nature and society, and the answers to both hopefully are not only based on assumptions but on knowledge about coevolutionary dynamics. And yet, the questions are put quite differently. From Kant's point of view, the questions refer to fundamentally different problems, problems that

---

[1]Cf. e.g. Keil (1987), who claims that Descartes was the first to make the transition from the primacy of ontology to the primacy of epistemology, which is a characteristic of modernity. For the entire process, cf. Gloy (1995, 163ff.).

are not equivalent and cannot really be compared. Whereas the first question clearly, even explicitly raises an epistemological problem, the second one can easily be understood as a question about morals and about a guiding principle for moral action. Which means, according to Kant, that this question may not be discussed in the context of epistemology but of ethics. Obviously, the second question, too, has an inherent epistemological problem. For the news of the impending extinction of the Siberian tiger may well be false. But Kant strictly distinguishes between the world of knowledge and the world of action. We thus have to split the second question into two subquestions: First, how do we know whether the news about the imminent extinction of an animal species is in fact true? And second, how do we react to the imminent extinction of an animal species? Both subquestions are to be answered independently of each other. Which means that we cannot deduct a demand for action, or an ethical obligation, from the answer to the question whether the extinction of an animal is a fact. Already David Hume called such a deduction of normative statements from descriptive statements a fallacy (cf. Hume 1888, 469f.). Kant agrees with him (Kant 1899, 198f./B375; cf. also Höffe 1996, 206f.).

Let us first look at the epistemological problems. How do we know whether the climate change we observe today has been caused by human actions? How can we know whether the Siberian tiger will go extinct in the near future? Both questions deal with the causes of unique historical events. Hence, we cannot simply fall back on earlier experiences to answer them. This is true especially for the question about the extinction of the tiger in the Amur region in Siberia, as the Siberian tiger is not yet extinct but his extinction is expected to happen in the future. Yet the question whether the current climate changes are manmade—caused perhaps by industrialization and the concomitant energy consumption—confronts us with as well with a historically new event. Both questions present us with the task to make decisions that broaden our knowledge.

Are we even capable of making such decisions? If so, on what basis can we assume that our conclusions about something we cannot yet know (such as future events) are correct? These questions lead directly to the discussion of Hume and Kant and their views of the possibilities of knowledge. Both Hume and Kant begin with the cognitive human subject, not with the reality that is to be perceived and understood. But they arrive at different conclusions, mostly because of the different constructions of their respective epistemological theories.

Hume is an empiricist. He rejects as a potential source of knowledge everything that is metaphysical, speculative or merely theoretical. Instead, he builds his epistemological theory strictly empirically, that is, he builds it on experiences. All knowledge, he says, is derived solely from the perceptions of the senses. There are two categories of perceptions: impressions and ideas. Hume's "ideas" are not to be confused with the kind of ideas philosophy has talked about since the times of Plato. Hume's ideas do not exist outside of human experience; rather, they are conceived of because of experiences. They are pale reflections of our immediate perceptions, brought forth by memory or imagination.[2] Both kinds of perception can lead us to knowledge.

We can know, according to Hume, *matters of fact* and *relations of ideas* (cf. Hume 1910, 306ff.). For instance, we can intuitively sense air temperature (perception). As it is cooler in the morning than midday, and as temperatures change again in the evening, the concept of an average air temperature (idea) suggests itself. With the help of a measurement concept, we can establish as a fact the mean daytime temperature from several different perceptions of temperature. And we can calculate relations between various average daily temperatures. Relations of ideas are certain and apparent, according to Hume. This is how they differ from the facts that we perceive with our senses. This is how Hume describes his distinction between these two types of knowledge:

> That the sun will not rise tomorrow is no less intelligible a proposition, and implies no more contradiction than the affirmation, that it will rise. We should in vain, therefore, attempt to demonstrate its falsehood. Were it demonstratively false, it would imply a contradiction, and could never be distinctly conceived by the mind. (Hume 1910, 306)

---

[2]"In short, all the materials of thinking are derived either from our outward or inward sentiment: the mixture and composition of these belongs alone to the mind and will. Or, to express myself in philosophical language, all our ideas or more feeble perceptions are copies of our impressions or more lively ones" (Hume 1910, 300f.).

He, then, distinguishes between knowledge about matters of fact, and knowledge about relations of ideas—which we see in the laws of geometry, algebra, and arithmetic—with a simple question: Is it possible to imagine the contrary of a statement? For the statement "The sun rises" the answer is "yes". But for the statement "$1 + 1 = 2$" a contrary is not imaginable.

If what we perceive with our senses is not necessary to gain knowledge, then how can we gain knowledge at all? Hume says, we can verify statements about present facts intuitively with our senses and thus clarify whether the statements are true or false. But what about statements concerning something we cannot intuitively verify? Claims about the causes of global warming, for example, or the imminent extinction of an animal species. Or statements about whether the sun will rise tomorrow. How can we get reliable knowledge about something that is not actually present?

Hume finds the answer to this question in the various possibilities of how we arrive at conclusions. There is, he states, only one reliable way to conclude knowledge, and that is deduction. Deduction of a fact is based on the two prerequisites: First, on the knowledge of a rule that produces this fact, and second, on the knowledge that the starting conditions of the process of fact production have occurred. For example, we know the 'rule' that carbon dioxide is released when fossil fuels such as oil are burned. If we now also know that oil is in fact burned, we can conclude that carbon dioxide is indeed released.

However, we can only learn from deductive reasoning what we in principle already know. Deduction does not allow us to learn facts that are new and unknown to us.[3] But as we cannot organize all our

---

[3]This is why Hume notes: "The Indian prince, who refused to believe the first relations concerning the effects of frost, reasoned justly; and it naturally required very strong testimony to engage his assent to facts, that arose from a state of nature, with which he was unacquainted, and which bore so little analogy to those events, of which he had had constant and uniform experience. Though they were not contrary to his experience, they were not conformable to it" (Hume 1910, 378f.). And in a footnote he elaborates: "No Indian, it is evident, could have experienced that water did not freeze in cold climates. This is placing nature in a situation quite unknown to him; and it is impossible for him to tell à priori what will result from it. It is making a new experiment, the consequence of which is always uncertain" (Hume 1910, 379, fn. 2).

thinking and reasoning deductively, we find us, in terms of certain knowledge, in a tight spot. On the one hand, we need experiences to broaden our knowledge. On the other hand, certain knowledge about unknown facts or future events cannot be deduced from experience. No kind of reasoning can lead us with compelling necessity to facts that we are not aware of.[4]

But what about the principle of causality? We are, after all, capable of observing causal relationships such as the following: If humans rely for their energy production on fossil fuels, carbon dioxide is created. If large quantities of carbon dioxide are released, it changes the chemical composition of earth's atmosphere. If the chemical composition of our atmosphere changes, there will be consequences for the climate on earth. That we are capable of drawing such causal relations between facts or events, creates a certainty that, we may assume, is just as secure and reliable as deductive reasoning in mathematics.

Hume repudiates this view. He does not deny the fact that we can draw such relations. But the crucial question for him is: What is their status? They are, so Hume, only interpretations added onto the observation that event A follows after event B. There is no observable causality; neither can we see it, nor can we perceive it with our other senses. Hume denies that there is a priori knowledge—knowledge that is not deduced from our experience but exists before we make our experiences. This is why he sees causal relations simply as assumptions that we make out of habit. Only out of habit do we conclude that present facts are the result of past facts, and present facts, in turn, the causes for future facts. Our knowledge about absent facts relies on the—quite reasonable, Hume agrees—assumption of the principle of causality. But we cannot establish compellingly necessary relationships based on this assumption.

---

[4]It is this insight of Hume that made Popper (1994, 9; see also: Popper 1979, 27ff./85ff.) proclaim the Scottish philosopher as the "discoverer" of the so-called induction problem. Hampe (1997, 75), however, points out that Hume never used the term *induction*, even though he was familiar with it from the Aristotelian tradition. Therefore, Hume was not primarily concerned with the question of whether a general statement could be abstracted from single observations. Rather, we can assume that Hume, like other empiricists, did not attribute a specific knowledge-related function to abstraction (cf. Pape 1989, 192, esp. fn. 91).

It is simply not possible to certainly extrapolate from past experiences to an unknown future event.[5]

Hume then leaves us clueless as to how we can gain secure, objective knowledge about the imminent extinction of the Siberian tiger and the human impact on climate change. His empiristic epistemological theory has fundamentally undermined the belief that there can be certain knowledge. The classical view that we can distinguish between certain knowledge (episteme) and uncertain opinion (doxa) is hence called into question. All that remains is skepticism.

Today, skepticism—seen as a critical position against any kind of beliefs and against naive optimism—may seem particularly attractive. But it is doubtful whether a skeptical position can provide answers to the question of how knowledge is possible. Our everyday life practice and not least the scientific and technical achievements of our civilization show us that we can say more about the world than what our individual, subjective habits tell us.[6] Obviously, we can draw conclusions from our perceptions and develop reliable knowledge. Otherwise, we could not live in a house, let alone drive across a highway bridge. In this context, the fact that we need to revise and re-revise our knowledge at least partially, may even be considered an expression of the basic human capacity of cognition. Kant was convinced of it, and this conviction is

---

[5]"There are some causes, which are entirely uniform and constant in producing a particular effect; and no instance has ever yet been found of any failure or irregularity in their operation. Fire has always burned, and water suffocated every human creature: The production of motion by impulse and gravity is an universal law, which has hitherto admitted of no exception. But there are other causes, which have been found more irregular and uncertain; nor has rhubarb always proved a purge, or opium a soporific to every one, who has taken these medicines. It is true, when any cause fails of producing its usual effect, philosophers ascribe not this to any irregularity in nature; but suppose, that some secret causes, in the particular structure of parts, have prevented the operation. Our reasonings, however, and conclusions concerning the event are the same as if this principle had no place. Being determined by custom to transfer the past to the future, in all our inferences; where the past has been entirely regular and uniform, we expect the event with the greatest assurance, and leave no room for any contrary supposition. But where different effects have been found to follow from causes, which are to *appearance* exactly similar, all these various effects must occur to the mind in transferring the past to the future, and enter into our consideration, when we determine the probability of the event" (Hume 1910, 333).

[6]Whereas Hume (1910, 322) writes: "All inferences from experience, therefore, are effects of custom, not of reasoning."

what drove him in the debate with David Hume. His guiding argument was a simple deduction: If we cannot infer the validity of causal relationships and natural laws from experience, yet depend on them and in practice treat them as if they were general knowledge, then something is wrong with the empiricist epistemological theory.

Kant developed his own epistemological theory, the "critical transcendental philosophy". He writes: "I apply the term transcendental to all knowledge which is not so much occupied with objects as with the mode of our cognition of these objects, so far as this mode of cognition is possible *a priori*" (Kant 1899, 59f./B25).

He picks up a thread in the philosophical debate that Hume intentionally neglected, namely a priori knowledge. What we know, so Hume, is based on the sensory perceptions and experience.[7] And Kant concurs when he states "that all our knowledge begins with experience" (Kant 1899, 43/B1). But this temporal logic does not imply that all knowledge is only a posteriori knowledge and derived from experience:

> But, though all our knowledge begins with experience, it by no means follows that all arises out of experience. For, on the contrary, it is quite possible that our empirical knowledge is a compound of that which we receive through impressions, and that which the faculty of cognition supplies from itself (sensuous impressions giving merely the *occasion*), an addition which we cannot distinguish from the original element given by sense, till long practice has made us attentive to, and skilful in separating it. (Kant 1899, 43/B1f.)

Kant enters afresh the field of the speculative, of metaphysics, because empiricism cannot answer the question about what preconditions exist for experiential cognition, prior to experience. Even before Kant, this had been a concern of the so-called dogmatic philosophers who regarded only reason—and not experience—as an epistemic authority. Kant (1899, 19) explicitly refers to Christian Wolff, the

---

[7]"Now whatever is intelligible, and can be distinctly conceived, implies no contradiction, and can never be proved false by any demonstrative argument or abstract reasoning à priori" (Hume 1910, 314).

"greatest of all dogmatic philosophers". Wolff, so Kant, had shown by example, "the necessity of establishing fixed principles, of clearly defining our conceptions, and of subjecting our demonstrations to the most severe scrutiny, instead of rashly jumping at conclusions" (Kant 1899, 19). Kant never intends to revisit the speculative objects like "God" or "immortality" of this philosophical tradition. He is interested in the rules of thought and reasoning, which Wolff developed in an exemplary manner. By applying these rules, reason evolves. Therefore, Kant (1899, 117) builds upon the "transcendental philosophy of the ancients", to develop a "theory of a priori knowledge" (Vaihinger 1970, 467; translation J.J.), but one that is critical, not affirmative. In other words, Kant experiments with reason, to find out what can and what cannot be a presumed transcendental prerequisite of cognition. If we thus learn what makes cognition even possible, then we can use Kant today, too, to say something about which cognitive preconditions are necessary so that we can reliable judge climate change to be "human-caused" and the extinction of the Siberian tiger as "confirmed".

Searching for the prerequisites of cognition, Kant rephrases the epistemological dilemma Hume left us with. For one, he classifies all possible cognitions according to whether they were formed a priori the experience with an object, or a posteriori. And he differentiates between a synthetic or an analytical character of statements about cognitions (cf. Kant 1899, 49ff./B10ff.). When the predicate of a statement expresses something that is already contained in the subject of the statement, then he considers it an "analytic" statement. In the statement "sustainable lifestyles consume less energy", for example, the low consumption of energy is already contained in the subject of the sentence. For a high energy consumption is incompatible with the concept of a sustainable lifestyle. If, however, what the predicate expresses is not (covertly) contained in the subject, but a new aspect is added to the subject when it is put in relation to the predicate, then Kant talks about synthetic judgments. The statement "Climate change is caused by humans" is an example of a statement with a synthetic character. Kant also uses the term "explicative judgments" (Erläuterungsurteile) for analytic judgments, and the term "ampliative judgments" (Erweiterungsurteile) for synthetic judgments (cf. Kant 1899, 50ff./B11ff.).

This is the problem Kant tries to solve: How we can know something new about an object prior to our experiences with it? Phrased in his terms: How is a *synthetic* a priori *judgment* possible? Kant's answer to this question is the insight that geometry—and even more general—mathematics and the natural sciences are built precisely upon to such judgments. For the synthetic a priori judgments are the ones that establish the compellingly necessary character of knowledge.[8] To be sure, not all cognitions of these sciences are transcendental, "but only those through which we cognize that and how certain representations (intuitions or conceptions) are applied or are possible only *a priori*" (Kant 1899, 95/B80). Kant then evaluates the results of his examination of the possibility of synthetic a priori judgments in mathematics and the natural sciences, with regard to scientific metaphysics. It is then the task of scientific metaphysics to deal with a priori "ampliative judgments" in order to ensure the objective validity of knowledge.

But how can the creativity of the human mind and its capacity for imagination be applied in a way so that it becomes clear *how* humans are able to distinguish between correct and false statements about the world? We could find the causes of climate change in the doings of higher beings or aliens that we offended—our creative minds could easily come up with such a fantastical explanation. Why then do we not seriously consider it? Hume could point to the epistemic power of experience and to the normative power of habit and custom. We have no sensuous evidence of higher beings or aliens. That they are nevertheless occasionally used as an explanation, corresponds with a pre-Enlightenment view of knowledge.

Unlike Hume, Kant cannot simply refer to sensuous evidence. His solution to this problem is both unique and groundbreaking. The limits of metaphysical speculation lie in the cognitive subject itself. The world is not something beyond or outside of the subject or outside of consciousness. Rather, it is the subject itself that constitutes

---

[8]Kant writes: "What can be called *proper* science is only that whose certainty is apodictic; cognition that can contain mere empirical certainty is only *knowledge* improperly so-called" (Kant 2004, 4/AV).

what the world is—with its faculty of cognition, its forms of intuition and its categories. Only if we take into account the world-constituting function of the cognitive self, can we understand, according to Kant, "that we are capable of making statements about this universe that are both correct and outside of our experience" (Stegmüller 1989, xxviii; translation J.J.).

When working on the problem of cognition, Kant discovers subjectivity and its specific function.[9] On the other side—the object side—this corresponds with the distinction he makes in his epistemological theory between the "thing-in-itself" and the "phenomenon". The "thing-in-itself" refers to the unrecognizable aspect of an object. The recognizable and knowable aspects he calls "phenomenon". The things-in-themselves may, in fact, be perceived as something conceivable but, so Kant, "we can have no cognition of an object, as a thing in itself, but only as an object of sensible intuition, that is, as phenomenon" (Kant 1899, 15/BXXVI). With this distinction, Kant provides metaphysics with a criterion to limit pure speculation about principles and causes of reality. As the thing-in-itself cannot be experienced by the subject, it cannot be an object of synthetic a priori judgments. But this does not invalidate the concept. Rather, the "thing-in-itself" means "real objects" that affect us and excite our senses. Kant's position is made clear when he explains how it differs from idealist views:

---

[9]Kant describes this discovery in rather vivid terms: "It has hitherto been assumed that our cognition must conform to the objects; but all attempts to ascertain anything about these objects a priori, by means of conceptions, and thus to extend the range of our knowledge, have been rendered abortive by this assumption. Let us then make the experiment whether we may not be more successful in metaphysics, if we assume that the objects must conform to our cognition. This appears, at all events, to accord better with the possibility of our gaining the end we have in view, that is to say, of arriving at the cognition of objects a priori, of determining something with respect to these objects, before they are given to us. We here propose to do just what Copernicus did in attempting to explain the celestial movements. When he found that he could make no progress by assuming that all the heavenly bodies revolved round the spectator, he reversed the process, and tried the experiment of assuming that the spectator revolved, while the stars remained at rest" (Kant 1899, 19/BXVI).

Idealism consists in the claim that there are none other than thinking beings; the other things that we believe we perceive in intuition are only representations in thinking beings, to which in fact no object existing outside these beings corresponds. I say in opposition: There are things given to us as objects of our senses existing outside us, yet we know nothing of them as they may be in themselves, but are acquainted only with their appearances, that is, with the representations that they produce in us because they affect our senses. Accordingly, I, by all means, avow that there are bodies outside us, that is, things which, though completely unknown to us as to what they may be in themselves, we know through the representations which their influence on our sensibility provides for us, and to which we give the name of a body – which word therefore merely signifies the appearance of this object that is unknown to us but is nonetheless real. Can this be called idealism? It is the very opposite of it. (Kant 2004, 40/A63f.)

Kant then does not simply transfer the world—and with it climate change and the endangered Siberian tiger—into the subject. His theory does not deny empirical reality. But it enlightens us to the fact that the capacity to perceive reality lies within the subject. The subject perceives the world by means of two cognitive components which cannot be reduced to each other: sensibility and understanding. Sensibility is purely receptive perception; because of it objects appear to us as *empirical* objects. In other words, the thing-in-itself "affects our senses." It becomes an object of our empirical intuition when we define it in space and time. Both of these features are, according to Kant, forms of pure a priori intuition. In comparison, the second cognitive component, "understanding", means using concepts to make non-receptive references to an object. They are a priori references insofar, as the concepts are based on a priori categories of perception and thought. For we do categorize the diverse impressions given to us by sensuous intuition, according to their quantitative and qualitative aspects, for example. We wonder about their reality and whether they are possible or necessary. This is the very process of how an *object* emerges from diverse impressions.

Hence, cognitions are not seen as "objective" because they exist independently from a subject's perceptions.[10] Cognitions are "objective" insofar as the cognitive subject relies on elements that are free from experience. Kant needs to develop these elements for both his cognitive components, for *understanding* as much as for *sensibility*. Hence, he proceeds from the assumption that understanding (reason) has an effect on the object, because of the use of categories, such as causality. Equipped like this, we can now infer causal relations and gain certain, unquestionable knowledge about absent facts (such as the unknown causes of climate change and the imminent extinction of the Siberian tiger). For causality, according to Kant, is a category of our mind, which we do not have to infer from experience. Thus he delivers us from Hume's skepticism. His theory ensures us of the basic cognitive capacity of humans: We do, after all, possess certain knowledge about the world we live in—knowledge that goes beyond our individual habits of thought.

There remains one small drawback. Not even Kant's epistemological theory can ultimately guarantee that the conclusions we draw to deal with our ecological problems are the right ones. But his theory makes us understand what is problematic about the two questions I asked in the beginning of this chapter. Both raise classical a priori issues, which is one fundamental factor of uncertainty. The assumption that climate change may be caused by humans, as well as the fear that the Siberian tigers may go extinct in the near future, are based on observation data. Such data may be false, inapplicable or insufficient. But the more we rely in our cognitive process on elements of a priori knowledge, the more certain such statements become, according to Kant.

Another point should be emphasized as well. When answering questions like the second one, Kant suggests, another factor of uncertainty needs to be taken into account: humanity as a moral and not only a

---

[10]Kant absolutely rejects the possibility of such a realistic position in his epistemological theory. In his conception, the subject relates to the world solely via its a priori cognitive apparatus. The necessity or commonality of objects does not depend on the fact that they are objects of this cognitive process. Neither can they be deduced interpretatively from the objects in the cognitive process. Instead, Kant assumes that the objects themselves only emerge in the subject's cognitive process.

cognitive subject. It is precise because humans destroy the animal's habitat that the extinction of the Siberian tiger has been predicted as a future event. As mentioned above, Kant views human action as a distinct field of inquiry. Not cognitive-theoretical reasoning is at work here, but practical reason. For humans, as moral beings, Kant postulates free will as a transcendental idea. In it originates the ability to act autonomously. All predictions about future events in the relationship between nature and society thus always come with the reservation that humans can indeed change their actions through moral insight. Thus, notwithstanding the bleak predictions,[11] perhaps there still is hope for the Siberian tigers after all.

How the Epistemological Controversy About Reliable Knowledge Influences Sustainability Discourse

The epistemological theories of Kant and Hume make us aware of the difficulties sustainability discourse faces. The conflicted positions they outlined still define today how we can discuss and justify decisions about the sustainability of social developments. As a rule, we imagine an undefined future. And only objective insights can provide us with a certain knowledge base. But what does objective mean, anyway? And if there are objective insights, what do they tell us about how to shape an unknown future? Ultimately, we need to make subjective value judgments to advance sustainable social development (cf. e.g. Kemp and Martens 2007).

To insist, in this world of concepts, on the uncertainty of all knowledge is considered an accepted, even a respected position. Hume's skeptical doubt pointedly demonstrates the resulting attitude: In the end, all

---

[11]Inspired by Douglas Adams and Mark Carwardine's *Last Chance to See* (1990), the Chinese Yangtze River dolphin (lipotes vexillifer) was used as an example in the first notes on this chapter. Then I realized that this species was declared extinct in 2006.

knowledge is nothing but desperate belief, thus the underlying theme, a belief that is bound to fail. For nothing can be certain; all knowledge, whether from the natural or the social sciences, has to be viewed and treated with the caveat that they will potentially be refuted.

In debates about the environmental and natural conditions of modern society, this kind of attitude has gained support, especially since the failure of large-scale technologies. The Chernobyl disaster of 1986 irrevocably put "risk" as an important issue on the agenda. Niklas Luhmann and Ulrich Beck were the first to address the issue, and are partly responsible for initiating the debate.

Ulrich Beck's *Risk Society: Towards A New Modernity* was published in 1986, the English translation came out in 1992. After Chernobyl, everybody talked about the "risk society"; the term perfectly captured the general sense of unsettling unease. When Beck writes about "risk society", he does so not only in regards to society's use of nature. The ecological problems are just one of many areas of life where members of society are confronted with new unpredictable and uncontrollable modernization effects. What previously had been considered rational and dependable, has lost its validity, writes Beck. Family and religious affiliations, political allegiances, and union memberships are losing their traditional guiding function during the process of modernization.

Beck sees present-day, modern society in the middle of a comprehensive transformation process that emerged from a first thrust of modernization. Some of its unintended side effects have come into sharp conflict with the ideologies that initially established modernization. Thus, it is today considered naive and unrealistic to still believe, after Chernobyl, in steady progress through nuclear energy as the always better, cleaner, and safer source of energy. The industrial modernity, Beck states, has come to its end. A new era is about to begin, the era of the risk society. The following quote illustrates how Beck envisions the difference between the passing era and the one to come:

> The risks and hazards of today thus differ in an essential way from the superficially similar ones in the Middle Ages through the global nature of their threat (people, animals and plants) and through their *modern* causes. They are risks *of modernization*. They are a *wholesale product* of

industrialization, and are systematically intensified as it becomes global. (…) Risks such as those produced in the late modernity differ essentially from wealth. By risks I mean above all radioactivity, which completely evades human perceptive abilities, but also toxins and pollutants in the air, the water and foodstuff, together with the accompanying short- and long-term effects on plants, animals and people. They induce systematic and often *irreversible* harm, generally remain *invisible*, are based on *causal interpretations*, and thus initially only exist in terms of the (scientific or anti-scientific) *knowledge* about them. They can thus be changed, magnified, dramatized or minimized within knowledge, and to that extent they are particularly *open to social definition and construction*. (Beck 1992, 21–23)

In other words, according to Beck, Hume's epistemic skepticism has reached all of society. Now all depends on the members of society and how they respond to this situation. Beck at least hopes that we will finally move from a reflexive to a reflective organization of society (cf. Beck 1988, 1993; Beck et al. 1996).

The translation of Hume's doubt into social theory is even more intensified in Niklas Luhmann's work. For him, skepticism is not society's reaction to the collapse of the Enlightenment promise of progress. Rather, he interprets uncertainty in every respect as the fundamental problem of society per se. According to him, social structures, in general, emerge to create secure and dependable social relations. In his social theory, societies are seen as self-organizing orders of potential social relationships (cf. Luhmann 1997). They exist solely because of the fact that humans are not solitary but social beings, with no discernable causes or reasons for their existence. There is no plan or purpose behind changes in societies, nor are they the result of knowledge about the outside conditions of a society, like their embeddedness in the natural environment. Because for Luhmann, knowledge and science are themselves in a radical sense part of society as well (cf. Luhmann 1990). They are only a uniquely organized form, for society to perceive and thematize itself. Once constituted, social structures are fixed—science talks about truth, the economy about profitability and so on—without the possibility to affect fundamental changes.

What we see as knowledge and understanding, Luhmann considers the result of historically unique (and thus coincidental) social processes. For knowledge is construed by society, in accordance with prevailing structures. People in the Middle Ages believed that the sun moved around the earth. The social structures allowed no other interpretation. Whether signs contradicting this knowledge may lead to a transformation of established views, does not depend on the sign themselves. The crucial factor, so Luhmann, is how these signs are perceived by society.[12] Dead birds falling from the trees are not per se a reason to reconsider the large-scale application of pesticides in agriculture. Whether the connection is made, depends, if we follow Luhmann's logic, solely on the very limited possibilities offered by the prevailing social structures. For dead birds by themselves are not a social fact. Whether they will become a fact, that is, whether society will perceive them as a problem, is—Luhmann claims—completely random. It could just as well turn out differently.

This fundamentally skeptical attitude towards rationality defines Luhmann's approach to ecological problems as well. He does not examine how modern society may adjust to ecological threats. Instead, he asks whether such an adjustment is at all possible. And while Beck, as we have seen, hopes for a second reflexive modernization, Luhmann, in his remarks, stays true to his skeptical constructivism.

The different concerns and styles of the two sociologists can at times distract from what they actually have in common. Both wish to enlighten society about itself: Beck when he draws attention to the dangers that industrial society poses to itself, and thus creates an awareness

---

[12]Luhmann considers this pointed argument the decisive contribution of his theory to the debate about ecological questions: "The theory of self-referential systems alone, however, has realized that the classical instruments of the acquisition of knowledge, namely deduction (logic) and causality (experience), are merely forms of simplifying the observation of observation. For social systems this means forms of simplifying self-observation. Methodologically, this means that the point of departure has to be the observation of self-observing systems and not the assumed ontologic of causality. In other words, one cannot avoid a decision about what counts as a cause and who is to be held responsible (Neither indications nor evidence and proof matter here, or they are considered of secondary importance, J. J.). It also means that morality and politics are overburdened by the unavoidability of this decision. The question then is how can this decision present itself so that the impression arises that it has not?" (Luhmann 1989, 9).

of the need of reflexive modernization; Luhmann when he enlightens the social sciences about itself and provides society with a theoretically defensible form of self-description, thus also changing society itself.[13]

"Self-enlightenment of society" might as well be the programmatic formula for the strands of sustainability research that draw attention to the cultural and social aspects in the perception of ecological problems. Our knowledge about the world and ourselves is always generated from and within social contexts. It depends on forms of thought and knowledge, on concepts, measuring methods etc., all of which were developed in historical processes.[14] But once this insight hypostatizes and the possibility of general knowledge is denied, skepticism as a figure of thought is inevitable.[15]

But cognitive skepticism is not all that can be expected from the sciences and their efforts for a sustainable social development. On the contrary, the natural sciences, in particular, try hard to establish reliable statements about causes and effects in the physical world. For example, we can now specify the dosage of pesticides that will kill a thrush. The results of experiments may fluctuate from one case to the next. But as soon as a sufficient number of individual cases has been studied, the generalized statement "A thrush will die if its body absorbed x milligram of substance y" can be viewed as reliable knowledge. In this context, "reliable knowledge" does not mean that the generalized rule will be valid and irrefutable for all times. The set-up of the studies could

---

[13]Luhmann's self-image is reminiscent of a spiritual teacher, whose insights are shared only with a chosen few. Thus he writes directly after the passage quoted above: "Radical theoretical positions of this kind lie far outside what social communication and ordinary consciousness accept today. Their consequences would require a rethinking whose results are unforeseeable. In any event, a period of slow and tedious development seems inescapable" (Luhmann 1989, 9).

[14]Articles in the field of "ecocriticism" (cf. e.g. Worster 1993; Glotfelty and Fromm 1996; Phillips 2003) are examples of this, also social-constructivist and ethno-methodological studies (cf. e.g. Sachs 1983; Hannigan 1996; Herrick and Jamieson 1995; Burningham 1998).

[15]When this insight is hypostatized into an (implicit or explicit) idea of reality, that is, into an ontology, we land right in the middle of the war of opinions between constructivists and realists. Absurd, over the top arguments can be found there. Like the one claiming that a fish only exists because of his socially constructed classification (cf. Tester 1991) or that the lactic acid ferment only became a reality through Pasteur's discovery (cf. Latour 1999). The achievements of the constructivist approach (cf. Burningham and Cooper 1999) are thus effectively undermined.

be flawed, no matter how often they have been repeated. Moreover, thrushes could adapt, gradually and to a limited degree, to the poisons. Still, such knowledge provides orientation, or what Kant calls an "empirical certainty". For him, truly secure knowledge is only possible in the field of cognitive-theoretical reasoning, in other words: in the material world, which the natural sciences explain with the help of a causal scheme. But empirical certainty is enough to provide us with basic points of orientation for our life choices and the shaping of society.

What then is the role of the social sciences within these ideas and contexts? The answer depends on what is seen as the function of the social sciences in society. As descriptive sciences, social sciences can, of course, study the causal relations of human actions and examine their causes. Depending on one's scientific stance, motives and decisions are considered possible causes for individual modes of action. If we can explain these causes in relation to the subjective intention of the actor, we can make statements about why humans acted like they did (cf. for an example, Weber 1980). But in light of Kant's views, these statements can only be considered valid or true for past actions. For the acting subject is free, and the field of practical reason is ruled by the principle of free agency. The causal relations of actions can be empirically observed, but this does not imply an inherently determinist view of culture and society. Culture and society are founded on freedom, and thus precisely on their independence from causality. Freedom, in turn, is defined as fundamentally different than regularities or structural patterns. Incidentally, Kant is not the only exponent of this view of reality. Whenever freedom of action is regarded as something fundamentally different from what we can describe as a rule or a structure, social science statements about people's life choices and about the structures determining social organization become useless. They cannot claim to be valid over time.[16]

---

[16]Game theory cannot make this claim either. Yes, alternate choices and strategies may be analyzed, with differing parameters and within various sets of meaning. But there is always something arbitrary about the resulting proposals. Games create closed worlds and they can be repeated. This is what constitutes their experimental character. But games cannot take into account that social processes are always defined by specific starting conditions nor can they account for the socialization of the players or agents.

Seen like this, it makes sense to assume that already tomorrow prac-
titioners of Chinese medicine will stop using tiger bones, and that the
inhabitants of Primorsky Krai in the border region of Russia, China,
and North Korea will decide to no longer destroy the habitat of the
Siberian tiger and start prosecuting poaching. In this view then, the
possibilities of the social sciences to make valid statements will inevitably collide with an anthropology pitting empirical certainties against the
human concept of self.

Yet, social scientists who want to make definitive statements about
society within the framework of Kant's (and others') views have another
choice: Social sciences do not need to be descriptive. They can be
devised as normative sciences. And as normative sciences, people's freedom of choice and action is their starting point. It is not considered
something that undermines the validity of scientific statements but is
seen as the foundation of human society, which needs to be maintained
(cf. Habermas 1981a, b; Dryzek 1987; of course also Beck 1992). In
this perspective, the contribution of social sciences to sustainability discourse is to encourage people in Primorsky Krai to develop an ecological awareness and thus change their actions. Social sciences, to put it
bluntly, are turned into specific forms of social pedagogy. As they reflect
on practice, they become themselves forms of practice. No epistemological function outside of practice can be claimed for them, and they have
no distinct research object of their own.

A summary of the considerations outlined above leaves us with a
sobering result: The contributions of social sciences to sustainability
discourse range from epistemic skepticism, on the one hand, to providing assistance, out of a sense of moral obligation, to ecological con-
sciousness-raising efforts, on the other hand. There are no other options
available within the cognitive frameworks from the eighteenth and
nineteenth centuries. As sustainability discourse constantly deals with
questions that address aspects of both the natural and social sciences,
one side is always underdeveloped.

But is there no alternative? The separation of nature—understood
as the realm of causality—and mind, the realm of subjectivity, has
long lost its rigor in research practice. This is in part owed to the

natural science and their reflections of the problem of knowledge.[17] Furthermore, the idea of the objectivity of the natural science itself has changed. Critical theory (cf. Habermas 1968), the so-called Starnberg School (cf. Böhme et al. 1972, 1973) as well as proponents of methodological constructivism (cf. Janich 1984) have rightly drawn attention to the fact that research in the natural science serves interests. It is never a pure, purposeless pursuit of knowledge, not even in the field of basic research. But it is not enough for sustainability research to note that the strict dichotomy between nature and mind has become brittle. The question is how *reliable* knowledge about the interdependencies of nature and society can be gained. Everyday experience shows that it is possible to gain knowledge we can depend upon: If experimentally determined amounts of pesticides are exceeded, dead birds will fall from the trees. A levee resists water pressure but it is breached when the pressure increases due to higher water levels. Yes, these are examples from natural and technological science contexts. But even our daily life practice shows us that we can determine social patterns and rules. If you see your colleagues at a new place of work throw tea bags, orange peels, and waste paper in one and the same garbage bin, you come to the conclusion that garbage is not usually separated here.[18]

---

[17]Kant's explanation for the objectivity of research in mathematics and the natural sciences was met with more criticism than agreement by those sciences themselves. Two developments were crucial for the rejection of Kant's ideas: first, the discovery of non-Euclidean geometries, and second, the work in modern physics that is not based on Newton's laws. Both developments disprove the two main assumptions behind Kant's assertion of the existence of synthetic a priori judgments: For Kant, Euclidean geometry was proof of their existence, and Newton's physics represented nothing less than Kant's ideal of a "pure"—that is, nonempirical—natural science, which is build in its entirety on synthetic a priori judgments. The developments in physics were caused by phenomena that did not fit Newton's view of the universe, especially electrical and optical phenomena (cf. Einstein and Infeld 1956, 51–85). Accordingly, Einstein, who used non-Euclidean geometry in the formulation of the general theory of relativity, attributed an empirical character to the axioms of geometry. Mathematicians such as Gottlob Frege, David Hilbert, Alfred N. Whitehead, and Bertrand Russell, however, generally attribute an analytical character to mathematics. Moreover, quantum mechanics questions the universality of explanations based on the classic models of causality. For an overview, cf. Höffe (1996, 61–65).

[18]Certainly, one could ask about the rules for garbage disposal. If the answer and the observation don't match up, a possible course of action is to develop a new question.

We could endlessly continue the conflicted debates that Hume and Kant introduced. But the need for a practical approach is urgent. It is time to remove the problem of how to produce knowledge at the interface of nature of society, from the debates about certainty and uncertainty. Epistemic skepticism is not interested in ecological questions and refuses to engage in working out tenable solutions to the problems. Sustainability research that aims at enforcing moral imperatives, is not interested in systematic research about the rules and structures in the nature–society relationship from the perspective of the social sciences, nor does it see the need to incorporate such research. We are looking for a new way to conceptualize the relationship between nature, society, and knowledge.

What would it mean to take up Kant's question about the capacity of cognition and apply it to ecological problems? Hume's insight that all knowledge has to be viewed with the caveat that it will potentially be refuted, could be a guiding principle. Knowledge has, after all, been observed from the position of people who live and act in concrete circumstances in space and time, and not from a position of all-knowing. We may be able to imagine such a position. But it is not a real option, just a dream of the poets of science. The other option we have is to inform sustainability discourse with reliable (not: certain) knowledge about the tangible interrelatedness of nature and society. The price is but a small sacrifice. We have to sacrifice the a priori determination that nature and *humanum*, or object and subject, are essentially different. As we will see in the following chapter, this sacrifice makes it possible for us to relate terms and concepts with sustainability discourse, terms, and concepts that allow us to produce reliable knowledge for a sustainable organization of society.

References

Beck, U. (1988). *Gegengifte. Die organisierte Unverantwortlichkeit*. Frankfurt am Main: Suhrkamp.

Beck, U. (1992 [1986]). *Risk Society: Towards a New Modernity* (M. Ritter, Trans.). London: Sage [originally published in German: Beck, U. (1986).

Risikogesellschaft. Auf dem Weg in eine andere Moderne. Frankfurt am Main: Suhrkamp].

Beck, U. (1993). *Die Erfindung des Politischen.* Frankfurt am Main: Suhrkamp.

Beck, U., Giddens, A., & Lash, S. (1996). *Reflexive Modernisierung. Eine Kontroverse.* Frankfurt am Main: Suhrkamp.

Böhme, G. (1997). Natur. In C. Wulff (Hg.), *Vom Menschen. Handbuch Historische Anthropologie* (pp. 92–116). Weinheim und Basel: Beltz.

Böhme, G., van den Daele, W., & Krohn, W. (1972). Alternativen in der Wissenschaft. *Zeitschrift für Soziologie, 1*(4), 302–316.

Böhme, G., van den Daele, W., & Krohn, W. (1973). Die Finalisierung der Wissenschaft. *Zeitschrift für Soziologie, 2*(2), 128–144.

Burningham, K. (1998). A Noisy Road or Noisy Resident?: A Demonstration of the Utility of Social Constructionism for Analysing Environmental Problems. *The Sociological Review, 46*(3), 536–563.

Burningham, K., & Cooper, G. (1999). Being Constructive: Social Constructionism and the Environment. *Sociology, 33*(2), 297–316.

Dryzek, J. S. (1987). *Rational Ecology, Environment and Political Economy.* Oxford: Basil Blackwell.

Einstein, A., & Infeld, L. (1956). *Die Evolution der Physik. Von Newton bis zur Quantentheorie.* Reinbek bei Hamburg: Rowohlt.

Esfeld, M. (2002). *Einführung in die Naturphilosophie.* Darmstadt: Wissenschaftliche Buchgesellschaft.

Glotfelty, C., & Fromm, H. (1996). *The Ecocriticism Reader: Landmarks in Literary Ecology.* Athens and London: University of Georgia Press.

Gloy, K. (1995). *Die Geschichte des wissenschaftlichen Denkens. Verständnis der Natur.* Beck: München.

Habermas, J. (1968). *Technik und Wissenschaft als "Ideologie".* Frankfurt am Main: Suhrkamp.

Habermas, J. (1981a). *Theorie des kommunikativen Handelns. Band 1: Handlungsrationalität und gesellschaftliche Rationalisierung.* Frankfurt am Main: Suhrkamp [also published in English: Habermas, J. (1984). *Theory of Communicative Action, Volume One: Reason and the Rationalization of Society* (T. A. McCarthy, Trans.). Boston, MA: Beacon Press].

Habermas, J. (1981b). *Theorie des kommunikativen Handelns. Band 2: Zur Kritik der funktionalistischen Vernunft.* Frankfurt am Main: Suhrkamp [also published in English: Habermas, J. (1987). *Theory of Communicative Action, Volume Two: Lifeworld and System: A Critique of Functionalist Reason* (T. A. McCarthy, Trans.). Boston, MA: Beacon Press].

Hampe, M. (1997). Unser Glaube an die Existenz abwesender Tatsachen. In J. Kulenkampff (Ed.), *David Hume: Eine Untersuchung über den menschlichen Verstand* (pp. 73–94). Berlin: Akademie-Verlag.

Hannigan, J. A. (1996). *Environmental Sociology. A Social Constructionist Perspective.* London: Routledge.

Herrick, C., & Jamieson, D. (1995). The Social Construction of Acid Rain. *Global Environmental Change, 5*(2), 105–112.

Höffe, O. (1996). *Immanuel Kant.* 4, durchges. Aufl., München: Beck.

Hume, D. (1888 [1739/1740]). *A Treatise of Human Nature.* Reprinted from the Original Edition in Three Volumes (L. A. Selby-Bigge, Ed.). Oxford: Clarendon Press.

Hume, D. (1910 [1748]). An Enquiry Concerning Human Understanding. In Eliot, C. W. (Ed.), *English Philosophers of the Seventeenth and Eighteenth Centuries: Locke, Berkeley, Hume* (Vol. 37, pp. 289–420). New York: P. F. Collier and Son.

Janich, P. (Ed.). (1984). *Methodische Philosophie: Beiträge zum Begründungsproblem der exakten Wissenschaften in Auseinandersetzung mit Hugo Dingler.* Mannheim: B.I.-Wissenschaftsverlag.

Kant, I. (1899 [1781/1787]). *The Critique of Pure Reason* (J. M. D. Meiklejohn, Trans.). New York: Willey.

Kant, I. (2004 [1786]). *Metaphysical Foundations of Natural Science* (M. Friedman, Ed. and Trans.). Cambridge: Cambridge University Press.

Keil, G. (1987). *Philosophiegeschichte II. Von der Aufklärung bis zur Gegenwart.* Stuttgart: Kohlhammer.

Kemp, R., & Martens, P. (2007). Sustainable Development: How to Manage Something That is Subjective and Never Can Be Achieved? *Sustainability: Science, Practice, & Policy, 3*(2), 1–10.

Latour, B. (1999). *Pandora's Hope: Essays on the Reality of Science Studies.* Cambridge, MA: Harvard University Press.

Luhmann, N. (1989). *Ecological Communication* (J. Bednarz, Trans.). Chicago: University of Chicago Press [originally published in German: Luhmann, N. (1986). *Ökologische Kommunikation. Kann die moderne Gesellschaft sich auf ökologische Gefährdungen einstellen?* Opladen: Westdeutscher Verlag].

Luhmann, N. (1990). *Die Wissenschaft der Gesellschaft.* Frankfurt am Main: Suhrkamp.

Luhmann, N. (1997). *Die Gesellschaft der Gesellschaft.* Frankfurt am Main: Suhrkamp [also published in English: Luhmann, N. (2012/2013). *Theory of Society.* Stanford: Stanford University Press].

Pape, H. (1989). *Erfahrung und Wirklichkeit als Zeichenprozeß. Charles S. Peirces Entwurf einer Spekulativen Grammatik des Seins.* Frankfurt am Main: Suhrkamp.

Phillips, D. (2003). *The Truth of Ecology: Nature, Culture, and Literature in America.* Oxford: Oxford University Press.

Popper, K. R. (1979 [1972]). *Objective Knowledge: An Evolutionary Approach.* Oxford: Clarendon Press.

Popper, K. R. (1994 [1934]). *Logik der Forschung.* Tübingen: Mohr [also published in English: Popper, K. R. (1965). *The Logic of Scientific Discovery.* New York: Harper & Row].

Rantis, K. (2004). *Geist und Natur. Von den Vorsokratikern zur Kritischen Theorie.* Darmstadt: Wissenschaftliche Buchgesellschaft.

Sachs, W. (1983). The Social Construction of Energy: A Chapter in the History of Scarcity. In *Schriftenreihe "Energie und Gesellschaft"*, Heft 22. Berlin: Technische Universität Berlin.

Stegmüller, W. (1989). *Hauptströmungen der Gegenwartsphilosophie. Eine kritische Einführung.* Band 1. Stuttgart: Alfred Kröner.

Tester, K. (1991). *Animals and Society: The Humanity of Animal Rights.* London: Routledge.

Vaihinger, H. (1970 [1881/1892]). *Kommentar zu Kants Kritik der reinen Vernunft* (2 vols.). Reprint of the second edition (1922) (R. Schmidt, Ed.). Aalen: Scientia.

Weber, M. (1980 [1921]). *Wirtschaft und Gesellschaft. Grundriss der verstehenden Soziologie.* Tübingen: Mohr.

Worster, D. (1993). *The Wealth of Nature: Environmental History and the Ecological Imagination.* New York: Oxford University Press.

3

Coevolutionary Science

The Difference Between Coevolutionary Science and the Established Sciences: A Scenario of the Future

"Our heads are round so our thoughts can change direction", wrote Francis Picabia in 1922. Nothing seems to be more difficult than to change one's habits and to look at things from a new angle. This truism also applies to sustainability discourse. Since the 1990s, there have been calls for integrated research of the ecological problems of modern societies, research that includes aspects from the natural as well as the social sciences. But despite the many attempts to newly define the nature–society relationship and to establish inter- and transdisciplinary research, the basic structures of thought have not fundamentally changed. Instead of searching for a new approach beyond old schisms, mechanical and causal-analytical versus holistic-hermeneutical argumentative models are very much alive. Either you establish an inductive or a deductive research logic. Either you research general laws or singular historical events.

© The Author(s) 2019
J. Jetzkowitz, *Co-Evolution of Nature and Society*,
https://doi.org/10.1007/978-3-319-96652-6_3

But when one aims at producing knowledge for sustainable social development, such simple schemes need to be avoided. The real question is: What makes coevolutionary science especially suited for this kind of knowledge production? Looking at a scenario of the future may be instructive in this regard, to sketch out an alternative epistemic practice, and to show how we evaluate possible actions differently when the social and cultural dimensions of perception, understanding and knowledge are taken into account.

Let us imagine the following scenario: It is the year 2040. The consequences of climate changes severely affect the living conditions of humans in large parts of the world. Every year, several ten thousands of people die because of typhoons in East Asia and hurricanes in the United States. Deserts and steppes are spreading and growing ever larger. Wildfires are raging across forest regions parched by drought. Twice in a decade, rivers in central Europe are overflowing their banks, causing what is termed "floods of the century". The ice in the polar and glacier regions has mostly melted. From the thawing permafrost soil huge amounts of methane gas escape into the atmosphere, increasing global warming. Confronted with this situation, a coalition of several countries decides to lower the global temperature by technical interventions. Several methods are experimentally tested. When potential side effects have been given their due consideration, a decision is made after four years: to slow down the rapid warming of the earth's climate, gigantic amounts of sulfur will be "injected" into the stratosphere. International diplomatic efforts to stop the project fail. Accompanied by protests of environmentalists, the project starts on July 4, 2039. Strictly supervised by the military and by scientists, great amounts of sulfur are lifted in high-altitude balloons from 10 to 50 km above the earth's surface. There the sulfur is burned in oxygen to produce sulfur dioxide (SO_2). In the chemical reaction, a layer of sulfate particles (aerosols) is created, reflecting solar radiation back to space and thus hopefully decreasing global warming.

This vision is less a science fiction scenario than it may appear. The idea to inject sulfur into the stratosphere is not new. It was developed by Russian climatologist Michail Budyko in the 1970s. In summer of 2006, the chemist Paul Crutzen revisited the idea and turned it into a realistic plan for crisis intervention, in case of "potentially more drastic global heating than expected".[1] It would cost between 25 and 50 billion dollars per year, Crutzen estimated, "or about '$25 or $50 per citizen in the richest countries of the world'".[2]

Paul Crutzen is not just anybody. He won the 1995 Nobel Prize, for his contributions to understanding the chemical processes causing the hole in the ozone layer. Because of his outstanding merits in climate change research, people are listening to what he has to say. And with his sulfur "injection" proposal he offers a tremendous promise: Ten thousands of people, who would have died of the consequences of climate change each year, will live, without the need for any kind of social change. Such a promise leaves a lasting impression; it cannot simply be brushed off with general reservations against what Crutzen has in mind. It does not matter that his proposed measures are called "engineering solutions", or that they are based on intellectually unsatisfying views of nature and society. The main point is their promise to save lives and to preserve the established status quo.

One can hardly argue against such an immense promise. It is like a religious promise of salvation: Act so and so in the present, and you will be rewarded with a blessed, carefree life in the future. And secular salvation is a promise for the here and now; it does not postpone the coming of this blessed, carefree life until after death, to the afterlife. There are many examples of such promises: the promise of salvation from typhus epidemics by preventive vaccinations, from agricultural pests by pesticides, from worldwide hunger by chemical fertilizers and green genetic engineering, from energy problems by nuclear power and hydrogen cells. Some of those promises have become reality; others are still waiting to be fulfilled. But the history of science has not always been a success story, and we should take care not to present new knowledge and new ideas in the same categories as redemptive religions. The idea of

---

[1]Paul Crutzen in the French science magazine *La Recherche*, quoted in: *Frankfurter Rundschau*, 4 July 2006, 1.

[2]Crutzen, quoted in: *Frankfurter Rundschau*, 4 July 2006, 1.

some kind of switch in nature, with which we can eliminate problems easily and for the good of the whole of humanity, seems extraordinarily naive when confronted with the reality of climate change and species extinction.

Proponents of large-scale geoengineering experiments do not generally believe that measures to counteract the impact of climate change will have no further consequences. Thus Crutzen points out that sulfur emissions already contribute to the acidification of the oceans. But he considers the increased pollution from additional emissions of the stratospheric sulfur "injections" as minor. The overall gain for the ecosystem is larger than its possible damage, he argues, even when further calculations are certainly necessary. And the measure he proposes would only be the last resort, the ultima ratio, if worse comes to worst. Ultimately, Crutzen is interested in a debate about "whether the climate system should be artificially influenced. (…) [This is] what we have to discuss, without any hysteria",[3] so Crutzen.[4]

But do we really have to? Such a discussion, after all, would only make sense if our search for knowledge was set in a space with no outside interests and purposes, in a space, too, with limitless access to research funds, time and creativity. Then we could, in the name of independent scholarly research, consider *all* possibilities to stop global warming—including those that without a doubt will have a destructive impact on other ecosystems. Only if this realm of pure, disinterested research existed, would it be useful to imagine a world where sulfur emissions protect the earth from solar radiation. Or a world where the allegedly carbon-neutral energy production from uranium does not emit harmful radioactivity, and where plants genetically engineered to

---

[3]Crutzen, quoted in: *Frankfurter Rundschau*, 4 July 2006, 1.

[4]Other climate researchers agree, e.g. Hans Joachim Schellnhuber, who also opposes "restrictions on thinking". Cf. *Frankfurter Rundschau*, 4 July 2006, 1.

adapt to climate change do not spread their genes to the environment. Considering everything we know today, we should no longer imagine such a world. It has been sufficiently proven that it does not further sustainable social development if knowledge and action are regarded as two categorically different variables. The examples of inappropriate scientific reductionism described above, and even more so the insight that all epistemic processes are embedded within social structures, motivate scientists to imagine an alternative of how nature, society, and knowledge are conceptually interrelated.

This raises complex issues: How can we make reliable statements about nature–society relationships when our knowing is itself a function of the society we live in, a society which in turn is fundamentally related to the natural environment which makes it possible? Can we develop an alternative epistemic practice, based on the assumption that the concepts and ideas we use to understand the world are implicitly connected to options for action? Can we develop a conceptual world which does not respond to complex circumstances with inappropriate simplification or abstraction?

My proposal for an alternative epistemic practice, to be delineated in the following chapters, must be capable of meeting high standards. The alternative epistemic practice has to lead towards a more adequate understanding of the interrelations between nature, society, and knowledge than Crutzen's proposal and the conceptual context that made his proposal possible. It has to account for an understanding of knowledge as an interminable process, yet still provide reliable, orienting knowledge for social practice. With my outline of the suggested epistemic practice, I hope to make plausible why coevolutionary science is a viable concept, to be used for developing a sustainable society, which is able to alter its structures to actively adapt to changing natural and environmental conditions. Proposals for geoengineered climate control hopefully will appear in a different light. For at the core of my argument are major discrepancies between the problems Crutzen perceives, and the suggested solution he offers up for debate, discrepancies that wholly invalidate his proposal.

How Coevolutionary Science Finds Its Problems: Perception as Explicatory Presupposition

When trying to fundamental rethink previous epistemic practices and develop alternative concepts, the first issue that needs to be addressed is the perception of ecological problems. How is it that we humans perceive a problem in the relationship of society and nature? How can we, with the knowledge that is available to us, deduce new, thus far unknown causes that may impact the climate or plant and animal species. Or to put it more generally: How does coevolutionary science find its subject?

Creativity as a Theme in the Philosophy of Science and Coevolutionary Research

Is it at all worth thinking about how new ideas and new knowledge emerge? This question is a contested issue in the philosophy of science. The creative aspects of scientific research are often deemed subjective, historically arbitrary or even irrational.

The most influential pioneer of these views is Hans Reichenbach, a logical empiricist and a proponent of a theory of science that takes physics as its prototypical model (cf. Reichenbach 1938). As far as the scientific research process is concerned, we can, according to Reichenbach, only comprehend the justification context of scientific theories. The philosopher of science has to examine how a theory is proven, the tests it has to undergo and the statements that are deducted from it, as well as their purview. In this view, it is of no interest under what circumstances a hypothesis has been formulated for the first time, or how the idea for an entire theory was discovered. These aspects are considered irrelevant for the quality of a scientific thesis. Historians, psychologists, and sociologists may illuminate the singular historical events leading to the discovery of a particular theory. But whether this theory is logical and consistent can in no way be decided by the con texts and circumstances of its origin.

In Reichenbach's view, it is also irrelevant how common theories about nature–society relationships in sustainability discourse emerged and how they arrive at always new descriptions of what constitutes an ecological problem. Without question, it is a historical event when the relationship of societies to the natural conditions of their existence is problematic, or becomes problematic. Perhaps the social situation in the United States in the 1950s contributed to it, or the intellectual prowess, the motifs and the energy of Rachel Carson and others. Historians may fight about the meaning of Carson's *Silent Spring* and the social relevance of the book.[5] But all of this does not make the nature–society relationship more or less problematic. And nothing of it has any bearing upon whether the knowledge and theories about this relationship are well or ill-conceived.

This distinction between the context of discovery and the context of justification seems overly plausible; after Reichenbach, it was adopted not only by the logical empiricists[6] but also by their critics.[7] However, adopting the distinction for a coevolutionary science would create problems, for the following three reasons: The first reason lies in the nature of the scientific practice itself. The discovery context and the justification context of a hypothesis or a theory cannot be differentiated quite as clearly as Reichenbach's distinction suggests. Studies in the sociology of science have shown that even in the ideal embodiment of experimental practice—in the research lab—the dividing lines between the two contexts are constantly obliterated (cf. Latour and Woolgar 1979; Knorr-Cetina 1981). The second reason for the rejection of the Reichenbach concept also comes from research in the sociology of science. Several studies demonstrated that there are patterns and structures in discovery

---

[5]In environmental history, all knowledge about the harmful impact of human action on nature is considered part of the tradition of sustainability and environmental studies. Cf. for a very striking example, William Kovarik's "Environmental History Timeline" (Kovak 2004) where he firmly warns against "dangerous myths" like the myth that one book—Rachel Carson's *Silent Spring*—"started all the uproar".

[6]Cf. e.g. Quine (1994, 49), who holds the opinion that the formation of hypotheses is not science, but the "art of science".

[7]Cf. e.g. Popper (1994, 7), who considers any kind of theorizing irrational as theories are developed intuitively.

processes (cf. e.g. Gigerenzer 1991), which imply that it is indeed possible to describe a certain rationality of discoveries. Consequently, it is inappropriate to cloak them in an aura of the mysterious, of genial intuition or of irrationality (cf. Danneberg 1989, 67f.; 1988). The third reason results from the perspective on, in general, the task or purpose of science that coevolutionary research has adopted as its own. Let me elaborate to explain what this means.

It is obvious that the purpose of science is to extend our knowledge of the world, that is, to produce knowledge. But how can science accomplish this task when only deductive conclusions are fully certain? This question already occupied Hume and Kant, but in the modern philosophy of science, it has been answered with the so-called hypothetico-deductive model of scientific inquiry (cf. Popper 1963). This model is supposed to further the developments of new theories about the world, and thus the extension of knowledge. Theory here means a cluster of statements that are internally consistent, have a (mostly) universal claim to validity and explain interrelated rules and structures. To justify a theory, hypotheses are deduced from potential observations and then verified by testing whether or not these observable facts do occur. A theory about the human causes of global warming would thus need to verify the following hypothesis: Whenever carbon dioxide enters the atmosphere, it results in global warming. With data taken from ice core samples from the Arctic and Antarctic, this hypothesis can be verified.

But the problems of sustainable social development can hardly be tested with this model. For while the hypothetico-deductive model is well-suited to verify theories about universal and necessary causal relations, its model specifications only allow reversible processes. When patterns and structures do not match this specification, they cannot be tested with the model. Strictly speaking, it is only for reversible processes that the starting conditions are irrelevant for the process sequence. The interrelations of nature and society, though, which are the focus of coevolutionary science, are not universal and not necessary; they are the results of historical developments that could, in theory, have gone other ways and arrived at different outcomes. As knowledge is part of the nature–society dynamic, the justification contexts of hypotheses and theories cannot be neglected—as Reichenbach wants to—for they influence greatly whether knowledge is established at all and when and how it gains acceptance.

Certainly, the conjecture that nature–society systems are irreversible is ultimately only this, conjecture. But it is a meaningful conjecture with enormous consequences for possible methodologies and the corresponding epistemic practices. If we ignore these consequences because we want the hypothetico-deductive model to be plausible, then we are left with only one conceptual possibility: We can propose a comprehensive determinism as metaphysics. Then the social developments which cause problems in the relationship between humans and nature, are also causally determined and nothing matters much anymore. But if we reject this conceptual possibility because it is inconsistent with the experiences that most research about coevolutions deals with, then we have to come up with other concepts of how humans get new ideas and produce new knowledge.

Let us go back to the question about the possibility of drawing scientific conclusions that extend our knowledge. If we want to avoid to basically postulate a fundamental demarcation between nature and society, we need to develop an idea of how new, historically singular constellations in the relationship of nature and society may emerge. To put it differently, coevolutionary research cannot limit itself to define universal and necessary patterns or natural laws that everywhere and at any time govern the course of the world. Instead, coevolutionary research needs to provide us with concepts and ideas about how new constellations may emerge and how we can explain the emergence of something new.

Creativity as the Implementation of Innovations?

We are looking for a theory of creativity, and a comprehensive one at that. It should be able to essentially describe how change and alteration take place, not as a variation of the same, but rather as the emergence of something new. This theory needs to be "comprehensive" in the sense that it does not only refer to changes of human thought and of society, but also to natural changes. For the observation, as Heraclitus phrased it, that no man can step into the same river twice, is not only the result of changes in the human perception of the river. The river, too, changes. What we are looking for then is, to borrow (and slightly adapt) the title of Bill Bryson's popular book,[8] a short theory about nearly everything.

Theories aiming to explain fundamental phenomena in a comprehensive way have a tendency to be overly abstract. Unfortunately, this high level of abstraction often goes hand in hand with trivial or incomprehensible statements, statements that the proponents of these theories dogmatically hail as the "new view of things". To avoid this, I like to first sound out the possibilities of how we can arrive at meaningful statements about the structure of creative processes.

What would happen if we discussed the creative aspects of knowledge, of society and of nature separately? We could initially focus on the question of how we recognize something new and extend our knowledge. The ontological debate about how we can even imagine a creative reality, a world in the making (Prigogine 1980; Prigogine and Stengers 1984; Whitehead 1979), could be provisionally suspended; and the question of what such an ontology means for action and for society (cf. Alexander 1987; Oevermann 1991; Joas 1992; Beckert 1997; Rammert 1997; Emirbayer and Mische 1998) could be postponed as well. Then all three theoretical components could be assessed separately from each other.

Let us look first at how the relevant academic disciplines explain the emergence of new knowledge. As described above, they examine the social implementation of new knowledge, not its emergence. Theories prevail—especially in the philosophical and sociological fields of the natural and technological sciences as well as in the so-called innovation studies—which only explain how and when specific innovations become implemented, focusing on the circumstances and the causes of the implementation (cf. e.g. Rogers 1962; Schumpeter 1912). The fact that new views and insights, new ideas and new knowledge emerge all the time, is usually taken for granted.

The textbook example of all implementation-of-innovation theories is *The Structure Is Scientific Revolution*, written by the philosopher

---

[8]The original title of Bryson's (2003) book is 'A short history of nearly everything'.

Thomas S. Kuhn (1962). Kuhn argues that scientific progress should not be seen as a linear accumulation of knowledge. Instead, scientific change takes place as a revolution within the scientific community. The established paradigm of scientific thought is radically and fundamentally replaced by a new paradigm, in what Kuhn calls a paradigm shift. Thus, the mechanistic world-view in the Newtonian tradition suddenly was regarded obsolete and was pushed aside by new views of relativity, even arbitrariness. Such revolutions, so Kuhn, take place as a generational change. The established generation of scientists is replaced by a succeeding generation that promotes the new paradigm. This sounds very plausible, especially to social scientists, as Kuhn's theory explains scientific change in regards to social factors. But the following issue quickly presents itself: Why does generation X of scientists break with the preceding generation and their concepts, whereas other generations show no interest in conceptual innovation? It is tempting to look for reasons in the new concepts themselves. But neither Kuhn nor other implementation-of-innovation theorists address this issue. Questions about how something new emerges, whether the new always comes from the old, whether it is even possible for something wholly new to come about etc. are seemingly of no interest to them.[9] We need a different approach if we want to learn anything substantial about the structure of creative processes.

Knowledge Extension by Purposeful Speculative Reasoning

Back to our initial questions: How is it that we perceive a problem in the relationship of society and nature, a problem which did not previously exist or which we only now perceive? How can we, with the knowledge that is available to us, investigate new, thus far unknown causes, which may, for instance, impact the climate or plant and animal species. And finally: How does coevolutionary science find its subject?

---

[9]Kuhn (1977) acknowledged these theoretical shortcomings of his theory.

To be able to formally describe the structure of creative epistemic processes, we need to rethink our view about logic. Logic cannot be only the theory of drawing correct conclusions. It cannot be limited to the possibilities of deductive reasoning, without adding something to the production of new knowledge.[10]

To overcome these limitations, studies in applied logic, especially in artificial intelligence research (cf. Simon 1973; Thagard 1988) and social research (cf. e.g. Oevermann 1991; Kelle 1994; Reichertz 1991, 2003), have adopted a broader understanding of logic as it is used in classical pragmatism. According to this broader understanding, logic, in general, is supposed to clarify the structures of thought and reasoning. And logic is supposed to take into account that we can only expand our existing knowledge if we accept the risk of a false conclusion.

The pioneer of this broader understanding of logic is Charles S. Peirce, the founder of pragmatism. Peirce differentiates between three basic types of reasoning which we use to organize our thinking about the world. Besides deduction, he postulates induction and abduction as two other possibilities of deductive reasoning. For Peirce, as much as for classical logic, deduction is the only truly safe mode of reasoning. In the transition from hypotheses to conclusions, the level of truth and the state of knowledge remain intact. However, induction and abduction are used, each in their own different way, to extend knowledge.

Induction extends the scope of knowledge: From a sample or a selection of instances, knowledge is induced about the entirety of instances. Statements about increased global warming, for example, rely on inductive reasoning. It is impossible to measure the temperature at every single place on Earth. Instead, the average annual temperature of the whole earth—the entirety of places, that is, of instances—is induced from the average annual temperature of representative locations. The reliability of inducing from a sample to the entirety of instances can be defined as "*probable* in the technical sense", as Peirce (1998, 287) writes, "i.e., … probable in such a sense that underwriters could safely make it the basis of business, however multitudinous the cases might be".

---

[10]As we saw above, already Hume and Kant struggled with these limitations.

It is different with *abduction*. This mode of reasoning can never, not even by degrees, be associated with the notion of *reliable* knowledge. In the discourse on logic, abduction has long been viewed as peripheral or irrelevant; the fact that abduction also means "kidnapping" certainly contributed to this.[11] Peirce attributes the term, which he introduced into the more recent logic debates, to Julius Pacius. Pacius translated the term "apagoge" in Aristotle's *Prior Analytics* (Book II, cp. 25, 69a) as "abduction" (cf. Peirce 1983, 90f.). What it factually means can best be described as a *purposeful speculation*, as a hypothesis that gives us a first idea of what may actually be the case.[12] Peirce (1986, 899f.) illustrates the function of the concept like this:

"Looking out of my window this lovely spring morning I see an azalea in full bloom. No, no! I do not see that; though that is the only way I can describe what I see. That is a proposition, a sentence, a fact; but what I perceive is not proposition, sentence, fact, but only an image, which I make intelligible in part by means of a statement of fact. This statement is abstract, but what I see is concrete. I perform an abduction when I so much as express in a sentence anything I see. The truth is that the whole fabric of our knowledge is one matted felt of pure hypothesis confirmed and refined by induction. Not the smallest advance can be made in knowledge beyond the stage of vacant staring, without making an abduction at every step".

Thus abductive reasoning is fundamental to any kind of knowledge. It is the ticket to the epistemic process, so to speak. Whenever we see something, we already make an assumption. We make judgments about

---

[11]In much of the English-language literature the term "inference to the best explanation" (Lipton 1991) is used instead, to describe the creative aspects of the scientific process. Lipton attributes the tradition of this aspect of the philosophical discussion to Peirce (cf. Lipton 1991, 58). But it is doubtful that Lipton engaged deeply with Peirce's concept of abduction, as Peirce's name is misspelled both in the text and the bibliography. Already Peirce himself realized that the concept of abduction would be hard to establish within logic discourse. Thus he writes in 1903 that the majority of logicians see abduction not as an argument but as a means. He stressed that this is an error of classification. Each of such means constitutes *ipso facto* an argument, since it tends towards a conclusion. The same authors would most likely accept an abduction in another part of their books as a generalization (cf. Peirce 1983, 96).

[12]"One comes to the conclusion that it is a question that may well be asked, and rightly so. This then is described as the forming of an explanatory hypothesis" (Peirce 1983, 95).

this "something" when we connect it, in our thoughts or with language, to another sign.[13] When we connect the sight of the flower in bloom to the word or the idea "azalea", we have already started an epistemic process. Whether the connection is correct or false, whether perhaps the flower is, in fact, a camellia, initially is of no importance. A fact has emerged (cf. Pape 1999).[14] The idea even goes beyond this: Any fact has emerged. What we know has not been deduced in a logical sense. For we do not learn anything new by deductive reasoning, as it does not reveal new knowledge to us. All our understanding and experience is built upon the fact that we know something by something else. It emerges from a combination of impressions with ideas, terms, concepts, which are all possible but not, and this is the point, necessary (cf. Pape 1994, 1999). Thus we understand the world we live in through assumptions. All our knowledge initially relies upon hypotheses, explaining something to us without compelling proof that the explanation is indeed adequate.

It is a provocative thought that corresponds with the uncertainty of all our knowledge. But how can we, under these conditions, arrive at reliable knowledge? Before I discuss this problem in the next chapter, I want to focus on what is implicit in the concept of abductive reasoning. Also, I want to clarify what abductive reasoning does not stand for.[15] To do so, we first need to analyze the "canonical form of abduction" (Kapitan 1994, 144), which Peirce (1934a, 117/CP 5.189) suggested in a series of lectures on pragmatism he held in 1903:

---

[13]Peirce does not fundamentally distinguish between thought and language. Both consist of signs and have, as have signs in general, a representative function. This is why Peirce says: "My language is the sum total of myself" (Peirce 1934b, 189/ CP 5.314).

[14]It is interesting to note here the similarities between Kant and Peirce's views. Kant (1899, 62) writes: "Thoughts without content are void; intuitions without conceptions, blind."

[15]Peirce kept elaborating and changing his concept of abduction, partly due to crises in his personal life, and partly because of the on-going development of pragmatism, which he had previously developed. This resulted in a wide potential for misunderstandings in the later reception of the concept, and led, especially in qualitative social research, to excessive expectations (cf. Reichertz 2003). The following remarks are based on the systematic reconstruction of Peirce's philosophical ideas (cf. Pape 1989, 2002) and his concept of abduction, as it has been defined by Helmut Pape (cf. Pape 1999).

The surprising fact, C, is observed;
But if A were true, C would be a matter of course,
Hence, there is reason to suspect that A is true.

This description of abductive reasoning raises two questions. The first addresses the surprising fact: Why is C surprising, anyway? What is so extraordinary about seeing an azalea in bloom? Something is surprising or extraordinary when it is not obvious and not a common, everyday experience. The azalea that Peirce saw on this spring morning in a year around 1900—and then described in the above passage—was certainly not the first azalea to ever be in bloom. We can, therefore, assume that the statement of something being surprising or extraordinary only makes sense in a specific context. It relates to a present moment, to a *hic et nunc*. Here and now something is surprising and in need of an explanation, and it is this here and now that the explanatory hypothesis refers to.

Defined like this, abductions as logical operations do not presuppose things that are obvious, habitual or every day.[16] Whether the azalea had not yet been in bloom yesterday, or whether Peirce just had not seen it bloom, is irrelevant in the moment when Peirce did notice it. There is no perception before the explanatory assumption. Therefore it is the starting point of thought and reasoning.[17]

On to the second question: What actually makes us relate the "surprising fact, C" to the suspicion "that A is true"? Let us look at the

---

[16]"Abductive hypotheses are entry points into the space of logical relations whether within a theory, or a system of logical relations between linguistic or mental representations and their cognitive role can only be appreciated if their singularity and context-dependence is taken into account" (Pape 1999, 250).

[17]It is helpful in this context to recall that Peirce developed the concept of abduction as an answer to Kant's question about the conditions of the possibility of thought. But unlike Kant, Peirce accepts the irreducibility of perceptions (cf. Pape 1989, 126). No matter how one views Peirce's answer to Kant's question—to reinterpret the concept of abduction on the basis of Piaget's insights, as Grunenberg suggests (2005), misses the point. For the distinction between logic and psychological principles (cf. Peirce 1934a, 118f./CP 5.192).

facts—the average annual global temperature increases, and the number of tigers in the Amur region decreases. And let us assume that we are surprised by these facts, as the deviation of climate data and population size lies outside of expected fluctuations. We can only extend our knowledge about these phenomena when we try to explain them. The causes cannot simply be inferred from the raw material, in this case the measured data. And they cannot be identified by induction, either. Instead they are unraveled by a rather speculative thought process. It is not pure speculation (what we usually would call "phantasms") because the hypotheses are postulated for a specific purpose. We speculate to explain the above average global warming and the slow extinction of the Siberian tiger. This purpose limits the speculative scope. The hypotheses need to be able to explain the existence of surprising facts. They need, therefore, to describe real possibilities.

This helps us clarify how coevolutionary science finds its subject. Proceeding from a surprising event (like global warming or the extinction of a species), it is assumed that this event developed from a problematic relationship between society and nature. In other words: To avoid future problems in the relationship between ecological and action systems, we are looking for objective knowledge about their interdependencies. Hence, the constitutive purpose of coevolutionary science is the production of knowledge for a sustainable social development.

How Coevolutionary Science Produces Knowledge: From Purpose to Objectivity

From purpose to objectivity? It runs counter our everyday intuition to assume that we could develop reliable, objective knowledge if we depend on subjective assessments. Can we turn *our* questions about the sustainability of social developments into the foundations of a scientific discipline? Should we not rather provide conservation and environmental protection efforts with explanations that lie outside of arbitrary individual judgments? When the gates are opened for subjective assessments, it is an invitation for vested interests, and it is precisely those vested interests that have caused problems for sustainability discourse in the first place. Whether it is climate policies or species protection, land or biodiversity management, building regulations or permits for the exploitation of natural resources—the intrinsic value of nature always falls behind economic interests and goal-oriented calculation.

These and similar questions about the constitutive rationale for coevolutionary science cannot be ignored; it is necessary to address them. It is very common, after all, to identify purpose-oriented actions with subjective, calculating and strategic modes of action. This view is also wide-spread in the social sciences. But as I show below, within the conceptual framework of an alternative epistemic practice, the subjectivity of an action is not a criterion for its assessment. Any action is connected to subjective assumptions. Any action puts (at least) one purpose into effect. But it is crucial to note which purpose is realized and whether this purpose has already been determined before an action, or only emerges in the performance of an action.

Action as Socially Conditioned Epistemic Process

Social sciences have a peculiar relationship to the term *purpose*. Many social scientists feel it is exaggerated and thus inappropriate to assume that all action is purpose-oriented. We are, after all, not always calculating our gain when we gift someone with a present or donate to a charity. And it is not only for strategic reasons that we are friendly to our fellow man and woman. In everyday life, we experience such behavior as an expression of a certain set of values, and thus should be regarded as such in sociological theories, as well.

When social scientists discuss how they can meet the various personal experiences of acting human beings, they do so using conceptual frameworks from the philosophy of the subject. These frameworks consider the ability to act as an exclusive quality of the subject which is confronted by immutable reality as its object. The differentiation between subject and object is regarded as fundamental, as constitutive. With Kant as an example, we have seen in Chapter 2 that this differentiation impacts our possibilities to define the relationship between nature and society. The same is true for the concept of action. That action is oriented towards purposes or goals, can—in terms of the philosophy of subjectinity—only mean one thing: that the subject conceived of these purposes, or chose them. For it is the perspective of the subject that is foundational, a subject that sets objectives and weighs the possible means to reach these objectives. But as humans are not constantly in the process of gauging and calculating, we identify additional modes of action that articulate the spontaneous-creative, the emotional and also value-oriented experiences of the acting subject.[18] The teleological character of actions is then put on the same level as calculating behavior.

But this is not necessarily so. We do not have to resort to the concept of the subject when we conceptualize the meaning of "purpose" and the role purposes have in perception, thought and action. Again, we can look to Peirce for inspiration. He painstakingly avoids using the differentiation of subject and objects as the starting point of his considerations. Thinking as well as reasoning are semiotic processes. Whenever someone sees something, thinks about something, refers to something, learns something, a third something is at play. This "Third"—a physiological stimulus, a verbal statement etc.—mediates between someone and something. Peirce considers this intermediary—he calls it "sign" or "representamen"—foundational and hence conceives of reality as constituted through relations (not through relata). It is a difficult thought that may be uncommon at first, but it has sweeping consequences.

Let us look at Peirce's paradigm to understand perception, thought, and action: the use of signs. People use signs in many different ways. We talk and phrase statements. We make gestures and express mood and attitude with our body language. Street signs tell us where to go. Diagrams explain how things are related. Images make us think or elicit memories. Whenever we use signs, a simultaneous process of understanding is implied. We have, after all, an idea about what we say, and about what the picture, which we have hung on the wall, tells us.

---

[18]Social science theories taking issue with this conceptual strategy base their opposition on the sources of empirism. This is true for approaches as varied as rational choice theory, as proposed by e.g. Hartmut Esser (1996), and the theory of autopoietic social systems by Niklas Luhmann (1984, 1997).

Which does not mean that all signs are clear and unambiguous. We misjudge gestures. And we misinterpret statements, sometimes even our own.

These various aspects illustrate that sign usage as paradigm connects communicative and epistemic functions. Signs represent something for someone. The concept "sign" can describe how we orient ourselves when dealing with the world. Thus we need a broadly conceived concept of "sign". It should center on a specific view of relationality. Anything *can* be related to something else. But it does not have to be. In fact, not everything is related to all else, even when this rigorous statement has become popular in discussions about the concept of ecology. It should be avoided, as it leads to compulsory relational constructions. A sign concept should not be based on the *facticity* of relations but on their *possibility*. But if something points to something else, and a factual relation truly exists, then there is a basis for a sign. All of this is summarized in the following definition by Peirce (1998, 272f.):

> A *Sign*, or *Representamen*, is a First which stands in such a genuine triadic relation to a Second, called its *Object*, as to be capable of determining a Third, called its *Interpretant*, to assume the same triadic relation to its Object in which it stands itself to the same Object. ... This means that the Interpretant is itself a Sign which determines a Sign of the same Object and so on endlessly.

Therefore, a sign is a relationship of three aspects: the sign, its object, and its potential interpretant. As noted above, the relationship of those three aspects is crucial. They are not thought of as things or constitutive components of reality. The sign's interpretant is not identical with the interpreting person or a subject. The concept is not limited to the description of activities of a consciousness. Global warming and the extinction of the Siberian tiger are also conceptualized as interpretants of a sign. People may, but they do not have to be conscious of them. If they *see* them, that is, if interpretants emerge in their consciousness, those interpretants for their part have to be interpreted.

Within this conceptual framework it is possible to conceive of purpose without a subject. Purposes are inherent in all signs, as we need signs to achieve something—optimal traffic regulation, to pass on a message or to express a feeling. Going beyond concrete contexts of sign usage, this is true for all signs: Their purpose is to represent something as an object. Hence, representation can be considered the general purpose of all signs.[19]

In this context it is worth noting that there is a temporal dimension immanent in the concept of purpose; there is always a futurist aspect to purpose orientation. Any sign usage in the here and now aims at representing something in the future. The interpretant of a sign is the last and, from the sign's viewpoint, always the future aspect of the triadic relation. The time gap time may be infinitesimal short, but it is always there. And, as will be seen, the fact that the interpretant can in turn be a sign for another interpretant establishes the normative dimension of this conceptual framework.

Let us return to the themes of sustainability discourses and coevolutionary science. Here, we can see clearly how purposes can help us reach objective assessments and establish binding rules of conduct. We start with the following question: Should we denounce the refusal of a government to acknowledge the human impact on global climate change and the extinction of species? It is a valid question. We cannot avoid that semiotic and reasoning processes are hypothetical, after all. Our definition of a sign even emphasizes how impossible it is to ever be entirely sure about an object. The data which climatologists examine are signs representing an object. We can understand them as records of the (actual) climate. But they are not the object itself. It is not possible to remove this difference, not even if immense amounts of data were collected. For the data needs to be interpreted, no matter how much of it we have. The interpretations of the data itself become "only" signs

---

[19]"The purpose of every sign is to express 'fact,' and by being joined with other signs, to approach as nearly as possible to determining an interpretant which would be the *perfect Truth*, the absolute Truth, and as such (at least, we may use this language) would be the very Universe" (Peirce 1976, 239). It is this correlation of purpose and sign that marks the difference between a linguistic understanding following Peirce and Jürgen Habermas' views. Habermas adheres to the dualistic conception of nature and society; he sees subject and object as fundamental ontological categories (cf. Habermas 1999). For Habermas' reception of Peirce (cf. Habermas 1971, 1995; Oehler 1995; Pape 1989, 63) (fn. 16) and 183 (fn. 86).

for further interpretations, but they never become the objects they represent.

Based on Peirce, a sophisticated concept of the object of a sign was developed in semiotics, a concept that is helpful for us. It differentiates between two aspects of the object of the sign, namely the "immediate object" and the "dynamic object". The immediate object is the one that is available to us in the communication about objects. The dynamic object is what suggests itself to us and starts sign chains.

This differentiation is less abstract than it may initially sound. Imagine a situation from a classic comic strip: Two people are walking across a bridge. While on the bridge, they perceive a sign that is a clear indication that the bridge is damaged. The sign could tell them, "Watch out! You are in danger!" But at least one of them misinterprets the sign. The reader infers this from the communication between the two. It is a situation that is often used in comic strips, in varied forms and other versions.

A great example is a panel from the comic "Häger the Horrible". The panel shows a view readers of the comic are familiar with: a simple wooden bridge stretches over a valley, with snow-tipped mountains in the background. Häger and his best friend Lucky Eddie walk across the bridge when Häger asks, "Did you just say twang?" The reader, but not the two characters who cross the bridge from left to right, can see that on the left side, where the bridge is attached to two posts, the rope just snapped with the sound "twang". The comical effect is achieved because the immediate and the dynamic object of the sound are worlds apart. When Häger asks Lucky Eddie the question, "Did you just say twang?", it is a possible interpretant. But it is unclear what the immediate object of the sign "twang" is supposed to be. Why should Lucky Eddie have said "twang"? Was it a joke? Häger's reaction, in any case, is very calm, unreasonably so, as the readers know. They can be delighted with the absurdity of Häger's false interpretation or be gleefully amused about what is about to happen. For Häger and Lucky Eddie will feel the consequences of the fact that the sign "twang" was not caused by Lucky Eddie but by the breaking of the rope that secures the bridge. The reasonable interpretant would be to hold on to the bridge to not slide down, or to start running to the other side before the bridge collapses.

The differentiation between immediate and dynamic object of a sign explains the semiotic situation as it presents itself to all acting human beings, including climatologists and biodiversity researchers. Their debate about climate and biodiversity data, about the methods of data analysis and modeling, are debates about an immediate object. How we understand climate change as an immediate object depends on specific constellations. Circumstances of time, the data situation, researchers' vested interests, available mathematical methods—all of these contribute to the fact that we can only ever understand the object of climate data from a particular point of view. Thus the conclusions drawn by climatologist can always be called into question, as new data makes us understand new aspects of climate development. The dynamics of the objects are not removed by the conclusions drawn.

But in this conceptual framework, the fact that all our interpretations and actions are always, on principle, unreliable, does not give cause for skepticism. It is true that we cannot close the ugly divide between immediate and dynamic object. We cannot take up the position of an all-knowing observer, place ourselves outside of semiotic processes and deal with the world from an omniscient point of view. But from our life practice, we can draw the hope that we can understand at least something, and not nothing.[20] We are, after all, making decisions all the time in our life, we experience that decisions were right and that our perceptions, thoughts, and actions were adequate, correct or successful

Such experiences can be explained by the fact that purposes are not only effective in our signs, but also in their objects. Semiotics, based on Peirce, rely on this vague hypothesis. It corresponds with the view that in our semiotic processes we are able to state something true about objects and draw correct conclusions. Put in the language of semiotics: We, in general, are able to develop final interpretants for objects. The final interpretant is the entire truth of a sign. It is a sign—whether in thought or in language—that fully corresponds with its object. To achieve such a correspondence of sign and object is ultimately the goal

---

[20]Peirce uses the expression "desperate forlorn hope" (1976, 343) to describe the presumption that the dynamic object may guide the semiotic process at least to a small degree.

of any sign usage. For if the purpose of a sign is to represent something, then we can assume that implicit in any sign usage is a claim to truth. In debate and in conflict we can (not: must) learn. Which means that we can change the purposes of our actions in order to bring them in correspondence with the purposes of the objects.

Obviously, this is also true for the interpretation of data about climate, biodiversity, and social change. Coevolutionary scientists can fall back on various hypotheses when they extrapolate from the data to problems in the nature–society relationship. When weighing the arguments, by using new methods, and in the analysis of new data, they discuss the immediate object of their research. The impetus to, for instance, perceive new data like the several catastrophic floods within too short time intervals, and to even develop the new methods and arguments—this impetus comes, hopefully, from the dynamic object.

Excursion: What We Can Learn from Astronomers: Kepler and the Self-Correcting Capacity of Science

It is a pious hope that reality is our teacher and helps us correct our false hypotheses and theories. But we can do more than hope. We can change our sign usage so that reality has more of a chance to be a corrective for our false views, allowing us—despite a general uncertainty—to gain reliable knowledge about climate change, species extinction and other problems in the nature–society relationship: From our abduction of a surprising fact we postulate a deduction, from it we draw the most daring conclusions, which are verified by induction. Only then the powers of self-correction are set free, which make this kind of scientific fixation of a conviction different from all other kinds (cf. Peirce 1934c).

The classic example of the self-correcting capacity of science comes from astronomy: Johannes Kepler's approach when he discovered the orbit of Mars around the Sun (cf. Peirce 1986, 394–396). Kepler (1571–1630) is well known today as a mathematician and astronomer. His groundbreaking laws of planetary motions and his achievements in the integral calculus and logarithms stand—for us rather incongruously—side by side with his Protestant piety and his position as the astrologer of Wallenstein and of three German Emperors. His mother was a healer knowledgeable in herbal lore, and between the years 1615 to 1621 she was suspected of witchcraft. Today, these biographical facts would discredit Kepler as a scientist. And yet he is considered a model of scientificity, precisely because he subjected his assumptions and judgments about the world to trials and experiments in an exemplary fashion. He was not interested in a superficial confirmation of his theories. Rather, he made bold predictions based on his theories, predictions that he again put to trial with no predetermined results in mind. The story of the discovery of the orbit of Mars illustrates which aspects of Kepler's approach are the most valuable for us.

Even as a child, his biographers tell us, Johannes Kepler was fascinated with the world of stars and planets (cf. Caspar 1995; Tiner 1977; Connor 2004). During his studies he learned about the heliocentric system of planetary motion developed by Nicolaus Copernicus, and adopted it. During this time, Ptolemy's model, which positioned Earth at the center of the planetary system, was still a serious competition for the new Copernican theory. For in both theories, the calculations and predictions of the positions of planets differed widely from the observations in the sky.

In 1594, Kepler worked as a professor of mathematics in Graz. The position required that he draw up an annual calendar with astrological and meteorological predictions. Hence, astronomy became a focus of his interests. A firm proponent of the Copernican worldview, he developed a cosmological theory in his first major work, "Mysterium Cosmographicum". He attempted to determine the relations of the planets towards each other, by describing each planetary orbit as a manifestation of one of the five Platonic solids (see Fig. 3.1). Today, this seems like an odd mix, trying to deduce the answer to a modern question from an antique body of knowledge. But the book made Kepler well-known in the contemporary scholarly world and brought him into contact with Tycho Brahe, a Danish astronomer.

Brahe was an empiricist. He owned astronomical instruments and was an experienced planetary and astronomical observer. In 1599, he started working for the German Emperor Rudolph II in Prague, as his imperial astronomer. Brahe was not fond of theories developed a

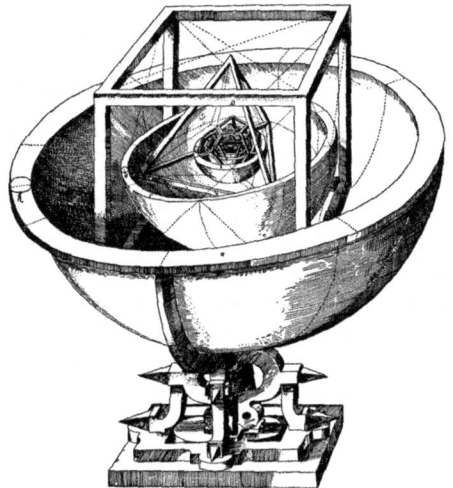

Fig. 3.1 Kepler's "Mysterium Cosmographicum"

priori,[21] but he brought Kepler to Prague as his assistant. Kepler's job was to mathematically process and systematize the huge amounts of observational data Brahe had collected. Brahe's goal was not to affirm the Copernican worldview; he was not interested in finding evidence for the Ptolemaic model, either. Brahe rather suspected the truth somewhere in the middle, and like others, he proposed a geoheliocentric view of the world.[22] According to this model (the Tychonic model) the planets revolve around the Sun, which in turn orbits Earth.

Kepler knew of the vast amount of data Brahe had accumulated. And Kepler needed such data if his theories about the harmony of the cosmos were not to remain pure speculation. To prove his theories, he

---

[21]"If the improvement of astronomy should rather be achieved, as you suggest, a priori with the help of the relations of those regular bodies, and not because of a posteriori facts gained by observation, then we will have to wait very long and maybe forever until somebody will be able to do so," Tycho Brahe writes in a letter from April 21, 1598, to the mathematician and astronomer Michael Maestlin, Kepler's mentor during his studies in Tübingen. Quoted from Lemcke (2002, 43).

[22]Voelkel (2001, 23) lists Tycho Brahe, Nicolas Reimers Ursus and Johannes Praetorius as proponents of the geoheliocentric model.

needed observational data, but it would take a lot of time, diligence, and patience to collect such amounts of data by himself. Kepler's situation was even more difficult as he had been nearsighted since childhood. However, once in Prague he quickly realized that Brahe was not willing to simply hand over his treasured data to his new assistant. Kepler was only given all the data pertaining to the orbit of Mars.

The situation was obvious then: Here the ambitious young researcher, there the renowned older scientist who stood in the way of free research with unlimited access to all available data. This situation changed suddenly and wholly unexpected when Brahe died of a renal colic in 1601, barely a year after Kepler had come to Prague.[23] Now Kepler could freely access all data collected by Brahe. He immediately started calculating the orbit of Mars but under the conditions of the Copernican model of the planetary system.

We have now come to the aspect of the story that is of interest for sustainability research: the approach Kepler took when he explored the orbit of Mars. Even though he was himself convinced of the Copernican view of a heliocentric cosmos, he still discussed the geocentric and geoheliocentric views of the world and evaluated which aspects could be better explained with these theories (cf. Krafft 2005). Kepler was not interested in using the data he inherited from Tycho Brahe to confirm the Copernican model as he had developed it in his book "Mysterium Cosmographicum". Such a deductive approach inevitably prioritizes the model and the hypotheses derived from it. Owed to his philosophical studies, Kepler was well aware of the danger of self-fulfilling hypotheses.[24] He did not set out to find confirmation for his theories, but he wanted to seriously test them. A (however well-justified) deductive selectivity when dealing with Brahe's data collection would have rather hindered than helped to achieve this goal.

---

[23]This story would make a great historical whodunit, perhaps? Cf. Gilder and Gilder (2004).

[24]At the beginning of his time as Tycho Brahe's assistant, Kepler had to work on a defense Brahe's against the attacks of another astronomer, Nicolaus Reimers Ursus. Trying to refute Ursus' arguments, Kepler was particularly concerned with the problem of formulating hypotheses (cf. Lemke 2002, 51–56).

What then is the alternative? An inductive approach seems feasible. Let us simply calculate a model using the data Brahe collected. And then let's just see which theory the model resembles most. As we have seen above, inductions are unreliable reasoning. And yet this approach is much more familiar to us today than to the people in Kepler's age, and not only because we are aware that our knowledge is always to a specific degree uncertain.[25] We also have computers at our disposal today, with computing capacities to process vast amounts of data in the blink of an eye. Calculations which can be done with a few mouse clicks today, Kepler had to perform in his head and by hand. Perhaps this is why Kepler never lost sight of something that we forget sometimes: that every date depends on—some would even say, is constructed by—a theory. Already Brahe's observations are based on theories that were translated into measuring techniques. But this is even more evident when we think of Kepler's problems while he was processing the data. Depending on whether the Sun or Earth was the center, whether the planets revolve around the Sun or Earth, or around both, the raw data needs to be differently transformed. Only with these transformations it becomes data that is actually useable. It must be emphasized that these transformations can under no circumstances be performed without thought, without particular assumptions, without theories and hypotheses. If we still mean to critically control these presuppositions, we cannot proceed purely inductively.

Kepler does not decide one way or another. His approach follows the above-mentioned sequence of abduction, deduction and induction, as Peirce first described it. We can call this the "abductive-critical" approach: accordingly, a hypothesis is neither deduced nor reasoned inductively. Rather, we arrive at a hypothesis on the basis of a reasoned, purpose-oriented speculation. We just have to think of our own lives where we constantly make assumptions, to be able to understand the world around us and all the events which may potentially happen as "facts to be expected". Besides these facts that we can expect, there are facts which will surprise us even within the context of

---

[25]Popper (1994) dates this insight all the back to David Hume's analysis of inductive reasoning.

our presuppositions. To explain these latter facts, we develop hypotheses. But we need to critical examine those, to be able to assess their real explanatory value.

Kepler's book "Astronomia Nova", in which he published his studies about the orbit of Mars, is an illustration of this process. He does not simply present the final results but recounts the story of an astronomical discovery, including all the errors that played a part in it (cf. Caspar 1937, 446).[26] Kepler and the other astronomers of his time were well aware of that fact that the orbit of Mars deviated from a circle. But what is the cause of this deviation? What holds planets in place and what moves them, if they are not—as Ptolemy believed—embedded in celestial spheres revolving around Earth? When Kepler starts to analyze Brahe's data to find answers to these and other questions, he makes several important presuppositions. Processing the data, he presupposes the Copernican worldview: Earth and the planets orbit around the Sun, and these movements need to be taken into account when calculating planetary positions. And Kepler presupposes that the Sun is the center of the planetary orbits, and not a point somewhat off-center. This last view was very popular in Kepler's time. Copernicus used it as a premise, to explain why planets did not move on a circular orbit. Kepler also had studied the ideas of magnetism, especially William Gilbert's book "De magnete", which had been published in London around 1600, and extended the physics of these ideas into the sky.

---

[26]The chosen form of presentation in "Astronomia Nova" is often seen as an indication of Kepler's underdeveloped sense of the stylistic requirements of scholarly texts. Thus, for instance, Koestler (1959, 314) writes: "Fortunately, [Kepler] did not cover up his tracks, as Copernicus, Galileo and Newton did, who confront us with the result of their labours, and keep us guessing how they arrived at it. Kepler was incapable of exposing his ideas methodically, text-book fashion; he had to describe them in the order they came to him, including all the errors, detours, and the traps into which he had fallen. The New Astronomy is written in an unacademic, bubbling baroque style, personal, intimate, and often exasperating. But it is a unique revelation of the ways in which the creative mind works." Whereas Stephenson (1987) shows that the book's rhetorical structure is by no means the result of stylistic inaptitude. Rather, Kepler conceived the book in exactly the form it was published, to assert his view of an astronomy that sees planetary motions as a result of physical forces (cf. Voelker 2001), for this point also Krafft (1973).

But even in the context of these ideas it is not explained what the orbit of Mars looks like. Kepler tests several hypotheses.[27] When he calculates an oval orbit, he notices that the form of Mars' path around the Sun looks more like a special type of oval, namely, like an ellipse. Kepler does not now conclude, "that the orbit is in fact an ellipse" (Peirce 1986, 394). But he starts to test this hypothesis, extrapolating predictions about different positions of Mars on the nighttime sky from the model "elliptical orbit". He then compares these predictions with the positions he calculated, using Brahe's observations. With this method, Kepler uses inductive examinations to test the idea of an elliptic orbit that he gained by abduction, after he translated it into a model from which deductions can be made. They are "serious" examinations as Kepler cannot know whether the positions he expects will actually coincide with the ones that were actually observed. He thus intends to sound out which statements he can infer from his hypothetic model.[28] His "Astronomia Nova" is an unabridged demonstration of how we can, guided by purposes, arrive at objective knowledge, when we use the potential for self-correction that is implicit in the sequence of abduction, deduction, and induction.

Dialogue and Debate as Means of Scientific Self-Correction

"Self-correction of one's own (or adopted as one's own) presuppositions instead of self-affirmation"—this simple formula explains what makes science scientific. Only by employing the principle of self-correction are

---

[27]"When Kepler discovered the ellipticity of the orbit of Mars, he met a surprising fact (the initial positions of the planet), then he had to choose between various geometrical curves, whose number was not infinite, however. Some previous assumptions about the regularity of the universe suggested to him that he had to look only for closed not transcendental curves (planets do not make random jumps and do not proceed by spirals or sine waves" (Eco 1983, 206f.; Cf. also Hanson 1965, 70–92).

[28]This is Peirce's (1986, 396) interpretation of Kepler's strategy: "He did not chose this test because he expected a favorable result. He did not know that this would be the case. He chose it because reason itself dictated him to perform this test. Continue on this course, and no theory will hold but the ones which are true."

we able to test whether the assumptions accompanying our sign usage are adequate, and only then we are able to change the purposes of our actions. The process, which Kepler demonstrates in the documentation of his own research, takes place today most often in an argumentative exchange within specific public spheres.

Climate research can help to illustrate how this process takes place. In 1896, Svante Arrhenius for the first time in the history of science published the hypothesis that a human-caused increase of carbon dioxide in the atmosphere will change the climate. Since then, the critical debate has continued and not come to an end (cf. Stehr and von Storch 1999; Viehöver 2003, 253–285). Originally, the debate was limited to scientific contexts and disciplines, but at least since the 1980s, it has reached the general public. This has been due primarily to the unusual increase of the average annual temperature of Earth's atmosphere and the oceans during the last century (cf. Smith and Reynolds 2005).[29] As for the causes of global warming, there are basically two views: The one side presumes, in the tradition of Arrhenius, that human-made greenhouse gases like carbon dioxide (produced and emitted since the beginning of industrialization) are responsible for the unusual increase of temperatures. The other side considers global warming a purely natural phenomenon. It differentiates between external factors, like fluctuations of solar radiations or volcano eruptions, and internal factors resulting from ocean circulation and its interchange with the atmosphere, or solely from atmospheric fluctuations.[30]

Today we have good reasons to assume that the drastic global warming is caused by the greenhouse gases produced by industrial

---

[29]There is no doubt that in the span of the last 100 years, Earth has grown unusually warmer. According to the Intergovernmental Panel on Climate Change (IPCC 2001, 2007) the mean global temperature increased by 0.6° Celsius (1° Fahrenheit) since the late nineteenth century. Since the 1970s, the increase has become particularly pronounced. The 1990s was the warmest decade since measuring of global temperatures began in 1860: 1998 was the warmest year, 2002 the second warmest, 2001 the third warmest, 1997 the fourth warmest and so on. Notably, global warming increased in (so far) two phases: from 1910 to 1945, and since 1976. There is no prominent increase in the period from 1946 to 1975.

[30]The IPCC (2001) assumes the net contribution to global temperature changes (solar radiation and volcano eruptions) during the last two, perhaps even four decades, was rather negative than positive.

civilization. This was not yet clearly understood 100 years ago. At the time it was known how the *effet de serre,* the greenhouse effect, worked. In 1824, Jean Baptiste Fourier had clarified why temperatures increase inside a greenhouse made of glass, as long as the sun is shining on it. The same principle causes the atmospheric heat accumulation on Earth, where a layer of gas takes on the function of the glass panels of a greenhouse: short-wave light particles are let through and heat the interior of the greenhouse, and Earth underneath the gas layer, respectively. As the glass panels, and the layer of gasses, are selectively transparent—they are more permeable from the outside to the inside than the other way around—the heat building up is not passed on to the environment in the same way. On the basis of the discovery of this principle, Arrhenius and, following him, Thomas Chamberlin quantified the importance of carbon dioxide as a greenhouse gas. In their assessment of how the massive release of carbon dioxide by industrialization would affect the temperatures on Earth, they both predicted that temperatures would rise.

A hundred years ago, not many people shared Arrhenius' and Chamberlin' view of human-caused global climate warming. Back then other theories were in vogue, proclaiming that climate changes depended on fluctuations of solar radiation, on sunspot cycles, volcano eruptions, and water vapor emissions—in short, on external or internal, *natural* factors.

Since the beginning of the 1970s, this situation has changed (cf. Viehöver 2003, 258–261). There is growing evidence for human-caused global warming. The accelerated increase of greenhouse gas concentration in the atmosphere during the last decades is well documented and easily verifiable, and this alone makes it doubtful that *only* natural factors could be the cause of global warming. The amount of carbon dioxide in the atmosphere, for example, has increased from 220 ppm in preindustrial times, to 365 ppm at the end of the twentieth century (cf. IPCC 2001, 38).

But caution is advised here: A simple comparison between the rise of the global mean temperature and the increase of the concentration of CO_2 in the atmosphere shows that these two facts are not directly linked. The phases of global warming do not correspond to the increase of carbon dioxide and other human-caused greenhouse gases. Moreover, the steady increase of atmospheric greenhouse gases does not explain the high-temperature fluctuations from one year to the next. In the scientific discussion about the causes of global warming, the central presupposition is rarely made explicit. Phrased as an "either-or" hypothesis, it is simply: Can humanity change the climate or not? The last 30 years of climate research basically are inductive examinations of statements derived from the deduction: Humanity is able to influence the climate. It is especially difficult to differentiate between internal, natural fluctuations in the global climate system, and external factors—human-caused or natural. Central for this problem has always been the reconstruction of the history of Earth's climate (cf. Fischer et al. 2004; Zachos et al. 2001).

After evaluating the range of studies and research results, the predominant view in climate research since about the mid-1990s has been that there exists an observable human influence on the global climate.[31] In the light of recent studies, the IPCC'S third report from 2001 attributes the causes for global warming in the last 50 years to human-caused greenhouse gas emissions.[32] Modelings also show that changes in solar radiation or volcanic activity had no significant influence on global warming during this period of time (cf. IPCC 2001). Since then, it has repeatedly been argued that the influence of solar radiation on climate development should be given more weight. But one look into available studies shows that solar radiation variations are too small to have contributed to the last phase of global warming (cf. Foukal et al. 2006). Furthermore, we find that, so far, not one of the studies conducted about the influence of solar radiation on climate history has taken surprising facts as its starting point. Consequently, there is no reason to

---

[31]The IPCC (1996, 5) concludes that "the balance of evidence suggests that there is a discernible human influence on global climate."

[32]"There is new and stronger evidence that most of the warming observed over the last 50 years is attributable to human activities" (IPCC 2001, 10).

expect that this factor will have a special explanatory potential when trying to explain global warming.

In summary and with the alternative ideas I outlined above in mind, it can be said that knowledge does not automatically become "objective" when the cognitive subject only relies on elements outside of their own experience. Only by critically examining their perceptions of the world and its phenomena, can people improve their chances to get a bit closer to the truth. At the same time, this view implies that in an alternative conceptual framework of cognitive practice, the search for knowledge has to be conceived of as a dialogic social process. This is true for debates about the causes of global warming in the context of the IPCC where scientists are engaging in dialogue about climate issues today. But these same scientists are also in dialogue with the ideas and assessments of Svante Arrhenius who developed the idea of human-caused global warming more than 100 years ago.

And this dialogue does not only reach from the present into the past. We cannot, after all, be certain that our research results will hold up to the test of future examinations. Any sign usage is accompanied by a regulative idea—the idea that it has to prove itself before an infinite community of past, present and future researchers (cf. Peirce 1934b, 86f./CP 5.311). This community is infinite, as neither temporal circumstances nor spatial nor social can be accepted as limits to the truth of the statement. Knowing that people in the ancient world envisioned the world as a flat disk, we can understand and explain various modes of speech and action common in antiquity. But this does not mean the statement "The Earth is a flat disk" was true in ancient times. The community consists of researchers insofar as anybody who uses signs knowingly or accidentally embarks on an experiment: to perceive, to conclude, to act and to inform somebody of something always holds the potential for miscommunication. Consequently, we rely upon continuously submitting our sign usage to inductive examinations. And if our tests suggest that we perceived or concluded incorrectly, we can change our sign usage. Which means nothing less than that we can learn from our mistakes and are able to redefine the objectives of our actions.

Who Benefits from Coevolutionary Science? Society as Sign Process and as Hypothesis

If human beings are basically capable of learning, it should be possible that we stop disturbing the ecological interrelations in ways that threaten our existence. But to be able to think in a meaningful way about how to make this possible, we first need to discuss the concept of "society". In the following, I put this concept in relation to two other concepts appearing in the course of the argument: "sociality" as a general term for all behavior relevant to any relations between living beings; and "humanity" in the sense of an infinite community of researchers. What does "society" mean in comparison to these two concepts?

Social Sciences and Humanity

"Humanity" is not a key term in the social sciences. It has become so obvious for social science studies to deal with strictly human behavior that animals had to practically be discovered as a new theme for social science theorizing (cf. Tovey 2003). Not many people today are aware of the historical debate about whether animal studies belong to the social sciences, especially to sociology (cf. König 1967; Portmann 1972; Tembrock 1997; Meléghy 2003). Social science was supposed to be a scientific discipline about human actions and behavior, according to Max Weber and others steeped in Kantian views. A clear and distinct dividing line was drawn, separating humans from animals. Animals do not perform actions, they have no free will, they are not capable of morality. Consequently, anything about animals that could be considered social aspects was delegated to biology. And to suggest that a sociology of *plants* could be part of the sociological discipline is completely out of the question.

But "humanity" is neither a biological nor a social term. What we call "humanity" is not identical with "society", nor with the whole of all societies. A biological species, too, is a very imprecise definition of the term. When the word "humanity" is used, it refers—we ascertain—to something other than a species. Using the word humanity means to express a conception of one's self (cf. Kamper 1997).

It is part of the human self-conception to conceive of humans as an exceptional biological species with one specific characteristic, namely, reflexive intelligence (cf. e.g. Mead 1967). Various philosophical and religious conceptions of the human being surmise that humanity possesses this characteristic exclusively. Yet any scientific examination of one's own convictions will show that the view of human beings as exceptional and distinct from other live forms—the plant and animal species on our planet—can only be regarded as abductive reasoning. Certainly, it cannot be treated as a fixed rule from which rights and duties can be derived. We do not need to assume the distinctiveness of humanity to see ourselves as human. Rather, it is a regulative idea[33]—a principle that has to prove its validity time and again: Are human beings truly distinct and essentially different from other kinds of living beings?

If the hypothetical character of "humanity" becomes the central factor of how we describe ourselves in the concert of all other living beings, then there is more at stake than is visible at first glance. For the question what constitutes a human being, and what makes them exceptional living beings, is now no longer solely the issue of anthropological positings based on religio-theological assertions. Also, the human self-conception is relieved of the task to universalize evolutionary biological and archaeological research and to philosophically account for it. As research situations can always change, after all, it is in general not methodologically advisable to base one's self-conception on scientific research results. Instead, only a future ideal can be at the core of what could distinguish

---

[33]As already mentioned above (cf. p. 19) the concept of a regulative idea goes back to Kant. Whereas constitutive ideas are at work in reality, regulative ideas (or principles) originate from reason alone and are not derived from experience. Kant clarifies this concept specifically in his letter to Karl Leonhard Reinhold from May 19, 1789: "Mathematics is the most excellent model for all synthetic use of reason, just because the intuitions *with which* mathematics confers objective reality upon its concepts are never lacking. In philosophy, however, and indeed, in theoretical knowledge, this demand for intuitions is one with which we cannot always sufficiently comply. When intuitions are lacking, we must be resigned to forgo the claim that our concepts have the status of cognitions of objects. We must admit that they are only Ideas, merely regulative principles for the use of reason directed toward objects given in intuition, objects that, however, can never be completely known in terms of their conditions" (Kant 1999, 306).

human beings from other living beings. Provided the human species lives up to this ideal, and that humanity proves to be an infinite community of researchers.

"Humanity" thus becomes a regulative idea, a limited concept which is wholly defined by a moment in the future. Only if the human species manages to organize its existence in an ecologically sustainable way, should we even consider to use the characteristic of reflexive intelligence as a criterion for what may distinguish humans from other living beings. It is easy to see that in this sense, humanity is clearly something other than society. When we talk about "society", we are not formulating a conception of ourselves as a biological species, neither as dogmatic positing nor as regulative idea. We refer to present, past, or even future forms of our coexistence with others.

The Difference Between Sociality and Society

Is "society" the same as sociality in general? This question, too, must be answered in the negative. But the reasons for this "no" are more complex if we refrain from making use of either dualistic views of nature vs. mind or anthropological positings (such as the exclusivity of reflexive intelligence).

For a definition of sociality, in general, I first turn to the ideas of the social psychologist George Herbert Mead. A classic pragmatist, he understands our lives to be socially constituted. How we perceive the world and the ways we influence and change it, always takes place within a social context, even when a single organism performs the act of perceiving, influencing, and changing. Like all classic pragmatists, Mead rejects the nature–mind dualism. Thus he is not forced to define the social context of life as only pertaining to human beings; he can develop a definition of sociality as a phenomenon effecting all of life. The following, longer quote poignantly summarizes his view:

"The behaviour of all living organisms has a basically social aspect: the fundamental biological or physiological impulses and needs which lie at the basis of all such behaviour—especially those of hunger and sex, those connected with nutrition and reproduction—are impulses and needs which, in the broadest sense, are social in character or have social implications, since they involve or require social situations and relations for their satisfaction by any given individual organism; and they thus constitute the foundation of all types or forms of social behaviour, however simple or complex, crude or highly organized, rudimentary or well developed. The experience and behaviour of the individual organism are always components of a larger social whole or process of experience and behaviour in which the individual organism—by virtue of the social character of the fundamental physiological impulses and needs which motivate and are expressed in its experience and behaviour—is necessarily implicated, even at the lowest evolutionary levels. There is no living organism of any kind whose nature or constitution is such that it could exist or maintain itself in complete isolation from all other living organisms, or such that certain relations to other living organisms (whether of its own or of other species)—relations which in the strict sense—do not play a necessary and indispensable part in its life" (Mead 1967, 227f.).

But as a phenomenon, sociality is not exhaustively described simply in terms of physiology. Mead clearly sees that there is another context which is effective across species. Hence, the second pole he uses to define sociality is *institutions*. What he means are habits and rules which govern the coexistence of organisms (no matter of which species) within a group. The single organism learns these habits and rules; it understands them to be the attitudes of the other organisms in the group.[34] Only because such institutions exist in a group of organisms, the individual can expect and predict the behavior of others. This makes it possible for the individual to act strategically.[35] Institutions here are not to be understood as purely "natural" products of biological evolution.

---

[34]"An institution is, after all, nothing but an organization of attitudes which we all carry in us, the organized attitudes of the others that control and determine conduct" (Mead 1967, 211).

[35]Here Mead as well as Durkheim side against the so-called social contract theory, today termed rational choice or rational action theory and similar approaches. Central for Mead is the understanding that social relations that go beyond purely physiological coordination are based on normative elements. These follow rules which make behavior expectable. At the same time, modes of behavior can vary. Durkheim's (1996, 256–286) opposition to social contract theory is based on his concept of society.

But neither are they purely "artificial" products created by a subjective authority—a spirit immanent in every form of life, or the human mind. Specific biological developments like e.g. the exponential growth of brain volume (cf. e.g. Esser 1996, 149–164) are necessary preconditions for the emergence of these institutions, but they can be examined and analyzed apart from them, so Mead.

But can these institutions really be viewed without taking into account their preconditions? This question is the focus of several strands of debate, reaching up into the present—the recent debate about sociobiology, for instance, the debate about organic and cultural definitions of gender, and last but not least the critique of Durkheim's paradigm— postulating the independence of sociological explanations—by human and social ecology. All three debates are part of a line of inquiry that started in the anthropological discussions of the 1920s: the question whether "nature or nurture" is the driving force of human behavior.[36]

Mead sees these two aspects as poles of a continuum of different forms of sociality. Different forms of sociality, he continues, allow also for different relationships between organisms and the environment. From an evolutionary point of view, the primary form of sociality is the adjustment of organisms to their living conditions by means of genetically coded information. A population receives indications about changing environmental conditions, and the subsequent generations react (subject to their capabilities) by implementing physiological changes in the organisms and thus adapting to the new living conditions. Forms of sociality where behavior is defined and controlled by institutions also allow organisms to adjust (subject to their capabilities) their behavior to changing living conditions. But what they need is a sign system that passes on information from brain to brain and thus makes communication possible.[37]

---

[36]The discovery of genetic coding has moved this debate from cultural anthropology to evolutionary biology. Cf. e.g. Ridley (2003).

[37]Psychobiologists (cf. e.g. Bonner 1980) differentiate between genetic and tradigenetic information reception. However, the underlying paradigm of information transfer does not pay attention to what media are used in this process. Hence, neither the inner logic of genes, nor the inner logic of culture/language is adequately considered.

Such sign systems can be established in many different ways. Usually, when asked what and how signs communicate, semiotics distinguishes between symbols, indices, and icons, which is a helpful distinction for us.[38] The terms represent the three ways in which signs can refer to objects. "Symbols" are, according to this distinction, signs that serve the understanding or the coordination of activities. They work due to a convention that a particular sign relates to a specific object. For example, each language is a symbolic system. Only if the signs are used by a community, and not just one person, can they signify something to somebody. Naturally not every member of a language community would have been asked if a sound should refer to something specific. For the most part, tacit recognition of what a symbol means emerges through socialization processes within a community. For the mere functioning of the symbols, it is irrelevant how horizons of expectation originated. The only thing that matters is that individuals share a set of expectations which determines to which object a certain sign refers.[39]

Not all signs function on the bases of conventions. As an "icon" and "index", semioticians discuss two other ways in which signs can relate to object. An "icon" is a sign that acts as a sign because of its similarity to an object. A diagram, for example, illustrates a structure of an organization or the course of a process like genetic coding. Photos depict creatures or situations. A map shows the order of streets and buildings of a city or maps other geographical conditions. Words like "bow-wow" or "cock-a-doodle-doo" imitate the sounds of dogs and roosters. Other terms such as "greenhouse effect", "silent spring" or "population pyramid" represent complex issues by referring to comparable but simpler ones. The principle of similarity is also at work in nature. For example, orchids mimic insects to attract insects. And the appearance of the harmless hoverflies resembles the aggressive and offensive wasps.

An index works differently. This sign is related directly to an object that causally creates it. Fever or pain is caused by disease. The

---

[38]Cf. e.g. Jetzkowitz (2010, 262f.).

[39]Referring to the quotation of Mead above (p. 97), it should be noted that the physiological aspect of sociality is conceived as a prerequisite for the emergence of symbolic sign systems, the institutional aspect is synonymous with symbols.

weathercock on the roof of a building indicates the direction of the wind if it can align itself with the wind. Dark clouds are signs of a coming rain and the moss growth on a tree refers to its weather side. A rising column of smoke indicates a fire. When someone calls "Fire!", it points to a fire and the dangers involved. Those who buy organic food or install solar panels on the roof of their home show an eco-conscious lifestyle. And with bark, a dog shows that it wants to be noticed.

The fact that we can communicate about the world with linguistic symbols because we learned *in specific situations* what these symbols stand for, is also what makes, generally speaking, for the difference between societies and socialities. As living beings learn situation-specifically to use the signs that are generally used in a community, a new dynamic development may emerge. This can result in three variables taking on a life of their own—the individual being (a), the community (b), and language (c).

(a) Individual beings may reflect on their sign usage. They can identify patterns, analyzing how they make gestures, phrase sentences, or employ other communicative conventions depending on the specific situation. One's own behavior thus becomes the object of a sign process, it is objectified. Which raises—at least theoretically—the question whether to continue the pattern or change it when next using the sign.

(b) Communities consist of living beings with a shared general level of expectation for symbolic-expressive behavior. Any situation-specific sign usage has the potential to express the difference of one community to other, similar groups. Whenever sign usage deviates from former conventions, and also a new symbol is added that depicts the concrete unity of the community, this difference becomes evident. When, for instance, a native from the Marburger Land in Northern Hesse pronounces the German "Heu" [hɔɪ] as "hieu", he does not only refer to *hay* but also indicates that he is from the village of Argenstein. His audience, provided they know the dialect, will either treat him as one of theirs or as a stranger—if they themselves pronounce "Heu" as "höe", "hiöe" or "höeueö".

(c) Whenever new personal or collective identities emerge, the symbolic sign system ultimately changes, as well. New symbols are added, older symbols are only rarely used or abandoned altogether. Perhaps they are still present in artifacts which now serve as evidence of a rich past. One can only speculate about their original meaning. As long as a symbolic system is able to integrate the various group identities—past and present as well as different local identities—the symbolic system will, metaphorically speaking, grow. It takes on a life of its own, not only independent from the individual living beings but also independent from any group. Such symbolic systems are generally called "culture".[40] What degree of independence a symbolic system can reach primarily depends upon how well it can be objectified and thus how well it can be recorded. Memories of past sign usage are retained differently when passed on from one mind to another than when they recorded in a book or saved on a computer chip.

Our initial question—what is the difference between society and sociality in general?—can now be answered in the contexts of these ideas: Societies can be formed when living beings situation-specifically use symbols or culture, and thus are able to modify their behavior (within one generation). When possible, communal constellations may become objects of signs. The distinctiveness, indeed the uniqueness of such a constellation can be symbolized and thus become generalized. Symbols expressing such communal experiences can be re-used in a subsequent situation and even be used to refer to new contexts.[41]

When a population, i.e. a reproductive community, passes on the distinct features of their sign usage through the succession of subsequent

---

[40]"Culture" therefore stands for symbols as a whole, not only for the ones which are in fact used by persons or communities. This distinction is crucial when trying to understand and explain social processes. Unsanctioned norms are part of culture and inform the potential for future social developments. Section 175 of the German Criminal Code, which made homosexual acts between men a crime, was no longer applied since the early 1970s. But until its deletion from the German Criminal Code in 1994 it was still possible for legislation to refer to this symbol.

[41]Several theories of cultural and social differentiation build on the development of different storage media of past sign usages. Cf. e.g. Parsons (1977); Habermas (1981a, b); Luhmann (1997).

generations, a society has been formed. In other words, a society emerges when a culture becomes institutionalized within a population.[42] With the continued usage of a specific set of symbols, it regulates interactions of individuals with and against each other and governs the coordination of activities. The individual being faces society in form of the generalized expectations of all other beings belonging to this society.[43]

Society Defined as a Community of Reflexive Sign Usage

The idea of society as a community of reflexive sign usage clarifies several important points: Unlike sociality in general, a society is a historical phenomenon. Societies, in the plural, each have a specific sociality. As communication traditions are established in populations, symbolic boundaries between populations of one species are created and a new independent aspect of reality emerges.[44] Of course, societies still reference their biotic and abiotic conditions; there *are* no societies without living and reproducing organisms. But they develop with a logic of their own, a logic which arises from the contextualization of symbolic signs. Bonds of solidarity that go beyond family relations are based upon identifying someone as same or as similar because they use the same symbols, because they relate to the same mythic founding father or the same divine first mother as I myself. This step—not tool use or the conduction of burials—marks the zero point where natural and social history part ways.[45]

---

[42]Hence, social change can be understood as the institutionalization of new patterns of culture, cf. Parsons (1964, 86); Jetzkowitz (1996, 26–53).

[43]In this way Mead's ideas about society are corrected and defined more precisely with a concept of indexicality. Cf. Pape (1995). Thus, when Mead talks about "society", more often than not "sociality" is the more appropriate term.

[44]Durkheim (1995, 109 and 203) views society as a reality *sui generis*.

[45]Elias (1998) rejects the idea of a "zero point" in the development of nature and society. He does, however, not distinguish between human sociality in general and specific societies in the sense described above.

Empirical research addresses the question whether this decoupling took place with species other than humans, or whether it could still take place in the future. The concept of society described above is open to any such surprises. As we couch our self-conception as a biological species not in terms of society but of humanity, it is conceivable that other biological species form societies as well (cf. e.g. Avital and Jablonka 2000; Witzany 2000). The discovery of the inner logic of human social relations is (of course) itself the result of a historical process. And as with any sign usage, the living conditions of the cognitive subject influence the epistemic processes.

Society as a concept was established at the end of the nineteenth century, at a time when national ideologies defined people's lives. It was mostly developed in France where the traditional legal and political orders had been dissolved in the wake of the Enlightenment and especially the French Revolution. Scholars like Saint-Simon, Comte, Spencer, Tönnies or Durkheim theorized about a comprehensive social context, but they also had practical goals in mind, namely the creation of new forms of coexistence. But can the idea of a comprehensive social system be applied universally? Is it not a typical late nineteenth-century European construct, quite unsuitable to grasp the realities of social life in earlier times and especially in other cultures (cf. Tenbruck 1981; Matthes 1985; Touraine 1986; Tenbruck 1989)?

When critics of the concept of society pose these questions, it is not to reject the underlying assumption that socially and culturally oriented living beings may develop with an inner logic of their own, apart from their biotic and abiotic conditions of existence. Such criticism is aimed at the idea of a comprehensive social context as a unified whole, as one organism (cf. e.g. Schwinn 2003). And certainly it is doubtful that societies, in the sense of a unified whole regulating whatever concerns its parts, ever existed or indeed can exist.

But this kind of skepticism does come cheap; it is aimed at an overblown interpretation of the metaphor of the social organism. If societies are nothing but the institutionalized cultural traditions of the respective present, they are never available as unified entities. Whoever claims differently, fails to recognize the true value of the organism metaphor: It emphasizes that the existence of societies—in the sense of a concrete albeit comprehensive context regulating the general rules of coexistence—attends to people's lives, at least as a hypothesis. Using this hypothesis, solidary relations are organized. Using this hypothesis, distinctions are made between who belongs and who does not.

During the history of humanity, the symbols societies use to organize solidary relations and to create affiliations and boundaries often changed (cf. e.g. Parsons 1966, 1971; Eder 1976; Habermas 1976; Habermas 1981a, b; Raeithel 1994; Giesen 1999). And the historical changes after 1945 may make it necessary to develop further concepts, demonstrating that the idea of a nationally constituted society has lost its persuasiveness and that other forms of society can emerge. Perhaps a new global horizon of expectation has emerged. We do not yet know the consequences and the impact this development will have on people's life. It is doubtful that "global" can be translated into "universal". With the help of signs, sociality can only ever be realized in a specific context. As discussed above, it is a regulative idea of social sciences that all people are connected within a comprehensive social system; it is not the near future.[46]

Societies can be understood as concrete albeit comprehensive social contexts, each providing its own specific conditions for the development of personal identities. A researcher in South Korea, dealing with the issue of how to reduce the production of greenhouse gases, will encounter different behavioral habits and different rules and laws than in Germany or France. These are the realities coevolutionary science has to consider and study. Ecological self-endangerment is not prevented by moral appeals. Which does not mean that debates about the purposes and objectives of human acts are superfluous. But such debates need to take into account the inner logic of societies, if they are not to be ignored.

---

[46]For a critique of social theories developed from the end of history, cf. Stark (2003).

How the Knowledge of Coevolutionary Science Takes Effect: Society and Its Future

Once we know how things stand, we are able to act. This is, in a nutshell, the classic definition of an ideal relationship between science and politics. "Classic" here means this way of thinking is a tradition guiding our actions. We assume as a matter of fact that this is true and also right: Know first what it is about, then act accordingly. We think like this on a small scale. For the big picture, it is phrased like this: Science objectively clarifies the actual facts, and politics uses this knowledge to shape society. This way of thinking seems so self-evident to us that any attempt to approach things differently needs to measure up to this model.

In social science discourse, knowledge is a long-standing and well-cultivated category. With it, social sciences have emancipated themselves from a discourse about truth dominated by philosophy. Knowledge can be observed, measured, and described, whereas truth can be either argued with or agreed upon. It cannot be denied that there is an ahistorical element in the definition of the concept of truth, which makes knowledge the more attractive category. By observing knowledge, one can perceive differences and changes over the course of history. The buzz word of the "knowledge society" even implies a current interpretation of contemporary social development. Other than in earlier societies, people in the globalized knowledge society today are eager to question, and if necessary to change their perception patterns and habitual actions. It makes sense then, in the context of the debate about the knowledge society, to also explore how planned social change takes place, in the correlation of knowledge production, action, and learning. The idea of planned social change is crucial for sustainability discourse. In the following, I focus on how the knowledge of coevolutionary science can bring forth a new sustainable society.[47] A comprehensive summary of the debate, however, is beyond the scope of this book.

---

[47]When social learning is discussed in sustainability discourse, it is usually with this question in mind (cf. Wals 2007).

I concentrate instead on the central antipodes: the idea of a society governed by knowledge, and the critique of this governance theory.

The Idea of Steering Theory: A Society Guided by Knowledge

It is an old idea that knowledge and action, or science and politics, are closely related. In the cultural history of Europe, the insight that knowledge is a necessary condition of virtuous acts, and that the wise act virtuous, is attributed to the Greek philosopher Socrates (cf. Platon 1990a, 561–567; Menon, 87d–89b). In his book, *The Republic* Plato bases a prominent, coherent conception of society on this insight. His attempt to found a community on the principle of justice culminates in the idea of the so-called philosopher king (cf. Platon 1990b, 445; Politeia, 473c–d). In an ideal state, Plato maintains, power should be in the hands of the wise and the knowledgeable. They have a natural tendency for virtuous activity; and in their souls, so Plato, the logical aspects are most pronounced. Fostered and perfected by education, these people are ultimately able to recognize the idea of the Good. It serves as their ideal when they rule the state and pass good laws. In the ideal state then, the philosopher kings govern the community in a way that allows everybody to do what befits them. Knowledge is only power in as much as virtue is knowledge; this is what Plato has Socrates say. In knowledge then, the exercise of power and ethical action coincide. The wise know what is befitting a person because only they can see the whole of it and recognize the meaning behind the phenomena of the world. Who would not want to entrust all political power into the hands of the wise?[48]

In a unique historical process, wisdom and knowledge started to become uncoupled during the European Renaissance. The driving force behind this process was the successes of empirical research. The combination of two practices was crucial, practices that were clearly separated

---

[48]This class of philosopher kings rules justly by definition and its decisions are above criticism— the elite of a totalitarian regime (cf. Popper 1950).

in the Middle Ages. The scholarliness of university teachers and humanists, which focused on logic and literature, merged with an interest in causal relations, experimentation and quantitative methods, which up to then had been the area of plebeian artists and artisans. From this combination emerged a new structure of thinking which was embodied in a new social role, the researcher or the scientist (cf. Whitehead 1925; Zilsel 1976). In this structure of thinking, knowledge is generated *not* by the exclusive display of eternally valid ideas allowing us to understand reality. Scientific knowledge is generated by dealing with the world experimentally, evaluating our experiences in a transparent manner and by a public discussion of the arguments in support of our view of things. This knowledge is different than the one available to the worthy members of a closed elite. Scientifically produced knowledge serves one end: to clarify and explain the causes and consequences of phenomena. A scientist's skills are to classify phenomena in a conceptual system and analyze them with a scientific method (cf. Burkholz 2008). The development of ideas about the meaning of the whole—of what is behind the phenomena of the world—is not the exclusive task of scientists. Of course, scientists are not and need not be immune to such questions. Johannes Kepler, for instance, could engage in empirical science and the lore of wisdom, in astronomy and astrology. Science is not free from worldviews, quite to the contrary: science is the discourse worldviews use to gain recognition. But worldviews which eschew discussion, claiming to be wisdom and to be valid without critical examination, have no business in science.

The scientific thought emerged during the Renaissance and was successively established all over Europe; it was professionalized in the role of the researcher and scientist. At the same time, this process led to a marginalization of wisdom traditions. At the end of the nineteenth century, hermetic and esoteric knowledge had been ostracized to a degree that it was no longer taken serious (cf. Vickers 1984; for Vickers' position, cf. Jobe 1986). Whatever could be stigmatized as "magical" or "occult" was disavowed. Knowledge was focused on the general and the necessary, the objective.

This process does not only pertain to knowledge of nature but also to the sciences concerned with human action and activities. Their historical development can be described less as a separation from the wisdom traditions; rather they emerged as a consequence of the disengagement of knowledge concerning a social order from morals and religion. This can be exemplified for sociology. Its founder, Auguste Comte, highlights this process of disengagement in his "law of three stages" (cf. Comte 1903). Comte's goal is to explore the forces ensuring order and dynamic, not only in nature but also in culture and society. Consequently, Comte initially talks about "social physics" instead of "sociology". The discipline, like all other sciences, is supposed to be more than a pure quest for knowledge, namely, it is supposed to be useful for the rational planning and shaping of society.[49] As a "human science", sociology is the leader of the group of positivist sciences (cf. also Elias 1970, 33). It can integrate the results of other sciences; it thus allows for politics to be based on science (cf. also Maus 1967, 21). Sociology, so Comte, is, therefore, a queen among the positivist sciences.

Even today social theorists share Plato and Comte's idea that the world in general, and society in particular, can be made governable by knowledge (and *is* governed by knowledge). The idea is of course no longer embodied in the ideal of the philosopher kings. We do, after all, live under democratic conditions, at least in societies with a European outlook or influenced by Europe. But the ideal of the steersman reveals a continuity of thought. It informs political self-presentations as much as sociological theorizing, focusing on the possibilities and limitations of political action as a question of political governance (cf. Mayntz 1980, 1983; Mayntz and Scharpf 1995).

And yet even the ideal of this changed idea is ancient. The image of the steersman belongs to the Christian tradition of the old church where the congregation saw itself as a ship sailing on the sea of history. But this ship is not blindly delivered to its fate, so the Christian myth. The ship has a steersman who is supposed to carry out God's will and

---

[49]"Thus the true positive spirit consists above all in seeing for the sake of foreseeing; in studying what *is*, in order to infer what *will be*, in accordance with the general dogma that natural laws are invariable," Comte writes (1903, 26).

guide the congregation. The term cybernetics is derived from the classical Greek word *kybernétes*, meaning "steersman". In the Christian tradition, cybernetics means the art or method of leading a congregation (cf. Schröer 1990). Understood as aiming at social facts, the concept of a steering theory was discovered long before electricity came from the socket, the information age could begin and Norbert Wiener would conceive of his ideas of cybernetics. Under secular and democratic conditions, the figure of the steersman embodies the ideal of a determined politician who is both knowledgeable and proactive. His actions are of course democratically legitimized. His knowledge is not derived from a special relationship to an omniscient God but is a result of his personal education and makes him amenable to scientific advice.

Thus the ideal; reality looks quite different. Even though Siberian tigers are threatened by extinction, illegal hunting of them persists, in order to provide a financially powerful segment of Chinese medicine with tiger bones. The governments of the 197 member states of the United Nations Framework Convention on Climate Change have not adopted effective measures against the unchecked release of carbon dioxide and other greenhouse gases, despite the fact that they have now been informed for years about the possible effects of these gases by the reports of the IPCC. The syndrome has been termed "resistances to advice" by those using the conceptual framework of steering theory. The causes behind the syndrome can be presented in detail. Russia suffers from enormous income disparities. Poachers living in the Amur region earn many times their annual household income with the sale of one dead tiger. Customs officials in the Russian-Chinese borderland have a hard time distinguishing the bones of tigers from those of other animals. Moreover, many still believe in the curative powers of medication produced from tiger parts even though their use was banned in traditional Chinese medicine in 1993. How various coalitions of interests time and again thwart an effective and sustainable climate protection policy can be illustrated by a decision made by German Chancellor Angela Merkel in the year 2007. At that time Germany held the presidency of the EU Council and at the same time was chair of the so-called G8 countries; it had proclaimed to officially use this influential position to work for climate control goals.

This proclamation did not, however, stop Angela Merkel from blocking an EU policy that would have required car makers to reduce fuel consumption. The automobile industry is one of the most important manufacturing branches in Germany, contributing substantially to the country's wealth.

We may react with moral indignation to such examples, an indication that we expect as a matter of course for people to not act against their better judgment. This is especially true when the people are politicians who in their official capacity make decisions not only for themselves but for a wider community. Perhaps it is our expectations and the ideas they are based upon which are at the root of the problem. Perhaps the idea of a scientifically enlightened governance of society by democratically legitimized politics is in need of revisions.

From Criticism of the Steering Theory to Skepticism About Society's Capacity to Learn: Luhmann's Theory of Autopoietic Social Systems

A fundamental critique of the idea of steering theory was formulated by Niklas Luhmann. I already introduced his social systems theory as an example for a position of epistemic skepticism (cf. p. 63ff.). In the discussion of his views about the role of politics (and its relationship to science) the problems of the ideal of the steersman become markedly apparent. The skeptical undertone of his societal theory gives reason to suspect that Luhmann generalizes from experiences of failed political planning. He studied law and at first pursued a career in public administration; his early publications predominantly deal with the causes of why political administration processes never achieve their targeted goals.

Luhmann's theory puts into sharp focus the autonomy of the social which I already discussed above. The way social systems function, they are impregnated against all information from the outside. They create their information themselves; Luhmann thus describes them as "autopoietic". It means that social systems reproduce themselves from within. The feelings and thoughts former US president Barack Obama may have had about the issue of "climate change" were irrelevant for his communications as president. Luhmann does not phrase this view as a guiding principle for presidents. He sees it as an accurate description of humans as psychic and also social beings. The US president's motions of consciousness define him as a psychic system. But consciousness is not what defines humans as social beings (cf. Luhmann 1984, 1997). Socially, the US president is a communicative event. When he gives a speech on the campaign trail, in front of the United Nations or to his team of advisors, each speech creates other communicative events. His speeches are interpreted, are commented on in the positive or negative, and they may be laughed at. His feelings and thoughts during his speeches, though, cannot be inferred from what he says, argues Luhmann. Social systems are simply different than psychic systems (cf. Luhmann 1992). The systems are closed and do not overlap, due to their different modes of operation. An element of a psychic system can never also be an element of a social system, nor vice versa.

Luhmann sees this principle of operatively closed systems also at work in modern society. According to him, modern society consists of diverse segments, each with its specific function representing the whole (cf. Luhmann 1989a, 1997). The economy supplies resources for society, religion supplies interpretations of meaning, the law supplies stabilization of behavior, medicine supplies health efforts and so on. Each segment can perform their highly specialized task because it structures communicative events according to its own criteria. Luhmann sees these criteria as binary codes. Each segment of modern society operates by its own code; the code allows it to tell whether a communicative event is relevant—or not relevant—for the performance of its function.

For instance, Luhmann identifies the criterion "payment/non-payment" as the economy's code (cf. Luhmann 1988, 1989a, 51–62). Once climate change has become an issue for society, it emerges as an economically relevant event as it is examined for its cost and profit factor, that is: as it is communicated in terms of payment or non-payment. A house owner may wonder whether or not he should insure his house against storm damage. In both cases, he communicates in the code of the economic system. If he regularly pays the insurance premium, the risk that he has to pay for eventual storm damage himself is reduced. But at the same time, the amount of money presently available to him is reduced. If the house owner decides against storm damage insurance, he can spend more money now. But at the same time, he risks that if there is storm damage he has to pay for the repairs of his house himself. No matter what the house owner decides, the economic system deciphers the communication with the binary code "payment/non-payment". It makes no difference to the economic system whether the house owner makes the decision because of an anxiety neurosis, because of calculations based on past experiences, or because of spontaneous decisions during the visit of an insurance sales representative.

The political system operates quite differently. It perceives house owners as politically active individuals, as voters perhaps, whose vote may depend upon a party's credibility concerning issues of climate policy, or it may depend upon promises to support house owners who suffered storm damage. The communication of house owners is deciphered by the political system with the binary code "power/non-power". Meaning, politics structures communicative events according to whether or not they may contribute to a rise to power or to the maintaining of power. Anything that is not decipherable with this code, so Luhmann, cannot be perceived by politics, and consequently politics cannot react to it (cf. Luhmann 1989a, 84–93, 2002).

That segments or subsystems of modern society, performing basic functions with specialized codes, operate fully independent of each other—this view is the background of Luhmann's radical critic of classical steering theory. The subsystems can after all only act towards their environment according to their code. Whether the scientific studies about climate change are correct, or whether it is true that the Siberian tiger is threatened by imminent extinction, is only of interest to the political system insofar as it may gain or lose power as a consequence. Politics cannot steer society, so Luhmann, as society is not steerable. Politics is neither at the top nor at the center of society but is only one subsystem among others. And as all subsystems have to operate according to their own inner logic, it is impossible that a coded communication may be connected to with a communication that operates under a different code.

In Luhmann's theory, all of this also applies to science (cf. Luhmann 1989a, 76–83; 1990). Science, too, is only one subsystem among others, without a specific guiding function for the whole of society. Science structures communicative events with the binary code "truth/nontruth". For the political system, it is risky to follow the scientific system (and for similar reasons, the legal system). An assertion like "The human-caused increase of atmospheric carbon dioxide contributes crucially to global warming" may be considered to be true today. But if tomorrow a scientist makes a coherent and convincing argument for the untruth of this statement, he is, according to the criteria of the scientific system, a successful scientist. He did, after all, refer to an earlier communicative event of science, following the rules of the scientific system. For a politician, however, who made his name as an advocate for climate protection, it would be a catastrophe if the idea of human-caused climate change needed to be fundamentally revised. He had to expect that his claims to governance would lose their basis, as voters would forever question his ability to adequately assess the situation. Instead of founding science-based politics, Luhmann's theory rather makes the case for politicians distancing themselves from science.

But this is not the only reason why Luhmann's social systems theory has been criticized from proponents of steering or—as they are named today—governance theories. They claim that of course "political governance" should not be understood to mean that "the state" or "politics" was purposefully guiding social development (cf. Mayntz 1987). Nevertheless, steering theorists insist that by passing administrative regulations or by other legal norms, society will be systematically changed (cf. Scharpf 1989). Ultimately, the history of environmental law, for instance, can be described as a history of political governance.[50] And this history has not exclusively been a history of failure, not even today. The reduction of sulfur emissions is an example demonstrating how legislation can—even though differently organized in different societies—purposefully influence society (cf. e.g. Münch and Lahusen 2001; Stark 1998). Luhmann's social systems theory, so the steering theorists, overstates conditions within societies.

---

[50]Cf. Radkau (2008), who published his overview of environmental history under the slogan "Nature and Power".

Meanwhile, it has been clarified in the discussion about social systems and governance theories that the steering efforts of a political system have an impact on other systems (cf. Luhmann 1989b; Scharpf 1989). Whether society and its subsystems will be changed by this impact, and whether this change will lead to a situation where societies no longer produce ecological crises—these questions are still debated.

Coevolutionary Science Between Steering Theory and Governance Skepticism

The controversy about steering theory helps us to get a clearer understanding of the steersman metaphor. Reduced to the act of steering, it can be interpreted as expressing a metaphysical conviction. So it is the mental capacity—the spirit, earlier scholars would have called it—which guides matter or rather, provides it with a direction, a goal. This basic structure depicts the central experiential component steering theorists aim at: Intentions can be causes of social change. Or put differently: Intentions give a direction to social change. If many people wish to avoid CO_2 emissions, fuel-saving cars, energy-efficient heating systems, solar and wind power stations etc. will gradually catch on.

Luhmann's doubts, in comparison, are plausible because of the multiple failures of measures which were or still are implemented, following this view. We all have experienced that sometimes even the best intentions cannot be realized. And so Luhmann builds his systems theory upon the view that intentions cannot govern communication at all because they are expressions of another autopoietic system mode, the psychic. Therefore it does not matter for social change whether people want to go green and try to reduce their CO_2 emissions. The wind blows wherever it pleases, and communication does the same. This succinctly summarizes the blueprint of Luhmann's theory of autopoietic systems. As a consequence, the knowledge of coevolutionary science cannot—from the point of view of this theory—take effect in a purposeful and systematical way. For knowledge in the form of cognitions from psychic systems cannot, so Luhmann, become part of communication systems. And knowledge in the form of scientifically coded communication cannot be made use of in politics, as political communication works with yet another code.[51]

Luhmann's conceptual world leaves few perspectives for coevolutionary science.[52] Because of this, and—as the steering theorists rightly point out—because laws for air pollution control and water quality protection are goals that have been successfully realized, it is worth thinking about other alternatives to put the knowledge of coevolutionary science to good use.

These alternatives become apparent in debates with positions from linguistic philosophy. Luhmann's theory seems to be clear about the significance of communication and language. But when Luhmann strictly separates intentionality from speech and communicative practice, he allows for only one of two possibilities to theorize the relationship between language and intention.[53] For alternately it can be assumed that language and intentionality are indeed related. Which then means that communicative practice—or more generally, sign usage—is constitutive of what we call mind, mental capacity or intentionality.

It is not my goal here to settle a fundamental controversy of linguistic philosophy. My discussion of this second possibility serves to find out what answers it may provide for the question about the effectiveness of the knowledge of coevolutionary science. Perhaps it allows for both possibilities. Perhaps one can, on the one hand, share with good reason the criticism of the naive belief that knowledge and insight quasi

---

[51]There is a third possibility from the point of view of Luhmann's systems theory, for the knowledge of coevolutionary science to take effect: the moral appeal. Knowing that the cultivation of mid-winter strawberries will result in the desertification of entire regions in Andalusia, consumers may decide to give up buying strawberries in winter to not become complicit in the development. But moral communication is, in Luhmann's systems theory, a parasite which only interferes with the operations of the subsystems of modern society (cf. Luhmann 1989a, 127–132).

[52]One perspective is Luhmann's response to criticism of his theory of autopoietic systems: Since the end of the 1980s he developed the concept of *structural coupling* (Luhmann 1997, 92–120).

[53]In linguistic philosophy this position was developed mainly by Daniel C. Dennett (1991) and advocates of intentionalist semantics like e.g. John R. Searle (1983).

automatically lead to changes in behavioral habits. On the other hand, one can perhaps also account for the experience that the (implicit and explicit) purposes of our actions of course intentionally influence social processes and developments.[54] In Luhmann's systems theory, the autonomy of the social, and the view that the social is purposefully governed by convictions, wishes, intentions, hopes, fears, and so on, do not go together. But they nevertheless may well belong together.

Let us try to develop another image of society. Let us try to connect the predominant self-organization of social subsystem (to which Luhmann rightly draws our attention) with the governability of society. Our starting point is the view that the production of knowledge and insights is a fundamental social process which cannot—not even under the conditions of modernity—be limited to one social subsystem, namely institutionalized science. To put it positively, knowledge production is an essential process for all social areas, as information provides orientation. I agree with Luhmann that the production of knowledge and insights is a communicative process, which is additionally defined by a specific temporal structure. But even more, I emphasize, we need to also consider the normative structure of knowledge production—to have available an adequate concept of this process, and to be able to critically assess actual processes of knowledge production (cf. Brandom 1994; Habermas 1981a, b).

---

[54] A simple combination of views from systems and action theory does not solve the problems. Ulrich Beck's concept of the risk society illustrates this point quite well. Beck uses theoretical elements from both conceptual worlds, but he does not systematically combine views of social order with intentional action. He refers to elements of social order theory, to show the problems brought forth by the institutions of contemporary society. The chances for reflexive modernization arise from these very problems. However, members of society need to realize these problems, act accordingly and resist the inner logic of the social subsystems. When he conceptualizes this inner logic, Beck relies on Luhmann's argumentation and adopts his concept of rigid systems. Consequently, Beck can only imagine that ecological problems will be solved by extensive cultural change, fundamentally transforming the structures of expectation—and with them the social and legal order (cf. Beck 1988). Thus Beck assumes historical periods and proclaims a second modernity (cf. Beck et al. 1996; Beck 1999). Such a hypothesis is irrefutable. A social event has either arrived or not yet arrived. The degrees of freedom to act within systemic orders—and thus the chances to change an order—cannot be substantially accounted for, as the concepts of action and order have not been set in relation to each other.

Normative Structures of Communication: Claiming Knowledge Creates Consequences

Communication is a process structured by norms. "Normative structure" implies the communicative process is structured by rules, and that it needs to be structured by rules. There is no other way. Communication only works with symbolic signs if their familiarity can be presumed. But not everyone who talks and communicates has to explicitly observe all the rules. All adult human beings experienced situations where they said something or replied to a question, without being fully aware of the impact of their words. If we take this insight in speech and communication serious, we arrive at the following conclusion: The view of communications as events connecting to earlier events—Luhmann speaks of sequences of "connecting communications"—unduly reduces the concept of communication. Each act of speech and communication follows implicit rules, and we can only understand the effects of language and communication when we understand these rules. If we make them explicit (cf. Brandom 1994), we can first reconstruct the meaning of what was said, and then, based on it, enter a well-founded argument about whether we will accept what was said, including implicitly expressed purposes.

This concept of communication is significant for the development of a concept of knowledge production processes. If we view knowledge production as a communicative process structured by time periods and norms, we can—on the basis of decision-oriented arguments—make a case for the idea of self-governed social fields.

As an illustration of what such an argument may look like, I use the discussion of Paul Crutzen's suggestion to decrease global warming by injecting sulfur dioxide into Earth's atmosphere. Our starting point here is the formal observation that we are in the same knowledge community as Crutzen. Because we partially share a language and a relationship to the world, we are able to understand his concerns. Additionally, we know—or can find out—that he has the reputation of someone who has knowledge. A layperson can only accept Crutzen's scientific beliefs and confirm him to be an expert in his field. His beliefs are regarded as truth, as long as we accept them to be true and subscribe to his views. With his contribution to the discussion about climate adaption, Crutzen makes—another formal observation—an assertion. He informs his audience about how things are in the world. Once declared, any assertion changes the relationship between the person who states something and the listeners who hear and acknowledge the assertion. There is a before and an after the assertion. Time is divided by the assertion as the changing element of the social fabric. The assertion also implements an asymmetrical relational structure. The person making the assertion also implicitly claims that he (or she) is allowed to make assertion. He makes clear: I know something and I share it with you. The listeners give the speaker recognition, they lend him authority. For Paul Crutzen, this does not seem to be a fundamentally new experience. He was already renowned as an expert before his assertion about the sulfur dioxide injections into the atmosphere. He had, after all, been awarded the Nobel Prize for his scientific research about the origin of the ozone hole. But his reputation does not relieve Crutzen of the duty to substantiate his argument. For as recipients of Crutzen's statement we participate in a "game of giving and asking for reasons".[55] Our strategic options depend upon our understanding of his assertion, and upon how we decipher the content of the statement as a network of premises and conclusions. In everyday communication processes we do this en passant; in empirical social research, we use complex methods to extrapolate the meaning of statements. Above we discussed how epistemic processes can never be finally concluded; this applies here, as well. Accordingly, we need to proceed from the purpose, to be able to decide how much effort is appropriate for each epistemic context. It does not take a complex reconstruction of motives and intentions to understand why Crutzen wants us to think about atmospheric sulfur injections: In an interview, he talks about the "greenhouse effect". We can conclude that he does not deny the existence of human influences on climate change. He presents his proposal with a calculation of costs, thus we can presume he

---

[55]Thus Sellars' (1997, 35) phrase for discursive occurrences, often quoted by Brandom (1994, 2001).

did not develop it on the spot. His proposal implies that our society is supposed to conduct an experiment with itself and its natural environment. There is per se nothing suspect about proposing an experiment. But the implications of Crutzen's real experiment stand out, especially when he refers to knowledge about ecological interrelations. Sulfur and sulfur compounds contribute to the acidification of terrestrial water resources. This fact, Crutzen claims, is not as bad as global warming. He thus suggests a kind of bartering of our possible futures. To secure the present power and property structures as well as the present territorial division of the planet for the future, we need to be ready to hazard the destructive consequences for all ecosystems. But only affluent societies can protect themselves against these consequences, and only they can replace services of the possibly impaired ecological system with technology. Thus from the perspective of a theory of sustainability addressing issues of social justice,[56] Crutzen's experiment has to be soundly rejected.

What is fundamentally wrong for one person, can be viewed in a very different light by another. Advocates of geo-engineering like Crutzen, Schellnhuber, and others may argue against the rejection of the sulfur injections experiment: In the history of humanity, such real experiments generated crucial technological and social innovations which made life safer and promoted the well-being of all. Already Joseph Schumpeter (1912) saw the creative elements in destruction. And have innovations in evolutionary history not always come with subsequent costs to which life then creatively adjusted?

The differences of the perspectives (likely) cannot be resolved. But the question about which assertion is now followed through need not now be decided purely arbitrarily or determined by the "right of the strong". According to Robert Brandom "far from precluding the possibility of conceptual objectivity, understanding the essentially social character of the discursive practice in which conceptual norms are implicit is just what makes such objectivity intelligible" (Brandom 1994, 55). The starting point here is that even in controversial communication

---

[56]Cf. e.g. Ott and Döring (2008).

processes, the conditions and consequences of assertions are in fact mutually assessed from each perspective. In doing so, the normative status of entitlement can be differentiated from the normative status of the commitment to a statement (cf. Brandom 1994, 115f.). The perception of the consequences of climate change entitles Crutzen to the statement that we need to think about injecting sulfur into Earth's atmosphere. Whereas he is committed by the assertion to define what a desirable future will look like. Is Crutzen entitled to this definition? The proposal, presented as the result of a simple, clear-cut consideration, would put—in an unforeseeable manner—an end to a multitude of real possibilities of development by depriving existing life forms not adjusted to high sulfur concentrations of their basis of life. Moreover, using this technology would bind the formation of future structures in the evolution of life permanently to an existence with acid water. And the resources of human societies would indefinitely be used for dealing with the consequences in other systems.

Crutzen cannot be entitled to this definition of the future, a fact that is evident not only to sustainability theorists. Once these implications of Crutzen's proposal have been made explicit, even advocates of geo-engineering strategies, who consider climate change the most pressing global problem, can no longer turn a blind eye to them. The fact that experiments with sulfur injections continue in the same behavioral logic that leads modern society into the ecological crisis, may be considered a secondary concern, as so far there has not emerged a feasible alternative behavioral logic for the shaping of nature–society relationships. And advocates of geo-engineering will not change their minds because Crutzen's proposal essentially concerns circumstances he has not researched, which renders his scientific reputation in this respect irrelevant. For they can always point out that Crutzen ultimately works for a good cause, namely the decrease of global warming. For people who consider securing the wealth of global society the most important task for the future, the dismissal of aspects of social justice and sustainability may take a backseat. But they cannot deny that Crutzen's proposal—if it is put into practice—does not remove the existential bond between human societies and ecological systems. The future may be contingent. We can thus wonder whether species could not adjust to the higher acidity of water, or whether even new species might come into being which are better adapted to the new environmental conditions. But a general possibility is not the same as the future (cf. Pilot 1972; Zimmerli 1997). Yes, what happens in the future is unknown from the point of view of the present. Yet the indeterminacy of the future is not absolute. Structures, which are inscribed in laws and behavioral habits, put limitations on which events are possible in the future and which are not. To hope for a creative power hidden in the alleged unknown, who will be able to form new life from sulfur, is an implicit assertion that, and geo-engineering advocates will agree, nobody is entitled to make. In other words, even from this perspective one can hardly look past the fact that a sulfur poisoning of ecosystems will have unforeseeable and incalculable results.

In this specific case, the game of giving and asking for reasons leads us—solely on the basis of already existing knowledge about the embeddedness of societies within ecological contexts—to the point where we have to consider Crutzen's proposal as untenable and thus over and done with. Not a ban on thinking, but a ban on action has to be declared. We need not conduct a real experiment to be able to understand its destructive results.

To be sure, such a clear-cut issue is the exception. For one, the knowledge that already exists about contexts and causal networks is barely made use of to think about possible coevolutions of society with related systems. And it should be noted that most often this knowledge does not exist. Let me underline once more that there is no certainty. It cannot be the objective of coevolutionary science to preordain future developments and changes. It cannot anticipate the future and it cannot define what societies should look like in the future. Its task is to generate knowledge about coevolutionary networks of relations, in order to better evaluate the pros and cons of decisions on social (and ecological) structures.

All in all, it can be stated: Knowledge in general—including the knowledge of coevolutionary science—has to be assessed in its social contexts. It takes effect when it is presented as a conviction, and when at the same the entitlement to present this conviction is socially accepted. Knowledge does not govern. It functions like a map by providing us with orientation in specific situations and showing us our next opportunities for action. We also can depict movements and changes on this map. Knowledge which is considered true changes when we pursue the irritations in our convictions, and then start explaining them in the light of new hypotheses. Coevolutionary science can only contribute to the shaping of the social future when it examines the factual results of existing and newly emerging opportunities for action, and assesses whether they guarantee or put an end to the open possibilities of development and codevelopment (cf. also Burger 2006).

Connecting Terminology and Ideas: The Concept of Reality in Coevolutionary Science

In the preceding chapter, I presented the outline of an alternative epistemic practice for coevolutionary science, a practice which does not categorically separate natural sciences and social sciences. I developed the view that people can understand the world they live in, even when they may never be ultimately certain that their interpretations of the world are correct. Like the idea of sustainability, the idea of objective knowledge is a regulative idea. It represents the desirable and at the same time utopian culmination of a development. Furthermore, I presumed that people change the world they live in. Not all changes take place casually and without intent, but some happen with purpose. Of course, not everything always goes according to our wishes and plans. Sometimes, purposeful actions fail, other times unexpected results occur beside the intended ones (cf. Merton 1936).

Still, we can assume that people have an idea of what they mean to achieve or produce with their actions. And they can compare this idea to the outcome of their actions. Do they correspond? Can something be improved? When they try again, the answers to these questions may help them to achieve a different and possibly a better outcome. This potential of the experimental structure is immanent in each action. Which does not mean that each action is performed under experimental or quasi-experimental conditions. It rather means that in the alternative framework sketched out here, cognition and action are not categorically different variables.[57]

Another aspect needs clarification before I turn to the question of how we can explore a world conceived as interrelated, perhaps coevolutionizing systems, and to possible consequences of this structure of thought. First, we need to consider how the various terms and ideas I suggested are connected. What view emerges in the synopsis? Does it differ in crucial issues from the view of the strict separation of nature and society? What about the concept of purpose which was introduced rather casually at a central position into this conceptual world? Do our views turn tautological because we keep drawing conclusions about something we already presumed? And finally, we need to clarify how these views fit with the metaphysical and philosophical hypotheses in recent debates, attempting to explain the relationship between nature and mind.

Of the Difficulties to Think Nature and Society Without Dualism

So far, coevolutionary science has developed ideas and structures of thought which avoid treating nature and mind (or society) as strictly separated worlds, spheres or systems. Instead, societies are conceptualized as concrete, symbolically constituted social contexts, embedded within nature. Their members, provided only with partial knowledge about their own conditions of existence, continuously influence their environment. How can people then, under these conditions, shape the nature–society relationship so to not cause unintentional side-effects which run counter to the vital interests of the human species?

It is hardly clear what these "vital interests of the human species" consist of. There is no authority to tell us what we need to do. Neither does nature teach us who or what we are and how we should act. As shown

---

[57]Hence we avoid what Whitehead (1925, 52) called the "fallacy of misplaced concreteness" and in this form criticized in Kant, namely that when cognition and action are differentiated analytically, they are turned into concrete units which then are treated as different categories.

above, already Hume and Kant agreed on the rejection of norm-settings which are legitimized with reference to nature. Such reasoning denies the experiences of humans as autonomous beings who are free in their choices and decisions. This still holds true today and including contemporary aspirations to justify social requirements with comparative behavioral research. The experience of autonomous action is the strongest legitimization of the nature–mind dualism.

But we need not reinvent the wheel to avoid this dualism. We can refer back to monist philosophical views. The term "monism" designates a view of reality which is based on only one fundamental principle and tries to understand everything from this principle alone. When, for instance, the physicist Stephen Hawking suggested searching for a "complete theory of the universe" (Hawking 2000, 225), a "theory of everything" from which, at least theoretically, all events of the universe could be derived, he, in fact, suggested a monist concept. In philosophies based upon contemporary developments in physics, monism has become quite popular.

The traditions of monist thought go far back to ancient Greece. Already the early nature philosophers were searching for the primal matter or the first principle, the *Archē* (cf. Keil 1985, 30). It became clear even then that all monist views come with a burden of reality. For they can only be developed based on the assumption that everything is a whole, a bigger context, a cosmos. As humans are part of the whole, the question arises how humans can claim to know the one principle of wholeness. Without metaphysical positings, it seems, humans cannot make this claim. Thus taking a monist position, one must expect severe criticism. For accepting a metaphysical positing as Archimedean point implies that one cannot self-critically revise one's convictions. This is not how science works.

Monists may argue that the nature–society dualism, too, is a metaphysical thesis. Separating the world into spheres of nature and mind means to posit a principle of reality, just as the monists do. But it is the strength of dualist metaphysics that it allows for the subjective as an independent variable. The subject makes experiences and can, as we have seen with Kant, respond critically to metaphysical positings. Monists lack such a critical authority. Their assumptions remain mired in metaphysics. We have to believe that the world is an integrated whole; that it can be traced back to one unifying principle. This belief is objective, it is real. It is hardly imaginable in monism that this belief could be shaken by anything leading to doubts or changes of this belief.

There remains one solution: The monists refrain from depicting the whole of reality as an underivable whole. And they develop a viable idea of what it is that mediates experience and reality, subject and object. We are looking for a term that can be valid as a basic principle of reality and as a basic principle of experience. The history of philosophy knows a myriad of approaches and conceptions of such a mediating authority. Already the dualist Descartes searched for and found it, in the form of a universal mathematics, a *mathesis universalis*. As a formalized universal science it was meant to allow an understanding of the world—in particular, the material world extended in space and time—in laws and variables. Leibniz adopted the idea of a *mathesis universalis* as part of his organicist views of nature and developed it into a general sign theory, which is also the foundation of metaphysics (cf. Mittelstraß and Schroeder-Heister 1997). The search for such a mediating authority has continued until the present-day philosophy of nature (cf. e.g. Weizsäcker 1971). But it can only relieve monism from its burden of reality when the metaphysical rationale is given up in favor of a general sign theory.

Only a theory which does not simply presume the sign concept as a unifying principle can count as serious contender for what we are looking for. The sign must be conceptualized as to be related to our experience and to be able to stand the test of our experience. Peirce's sign concept is set up like this (cf. Pape 1989). The result of a phenomenological analysis,[58] it allows us to define the universal categories of

---

[58]Peirce adopts the term *phenomenology* from Hegel, and he agrees with him "that it was the business of this science to bring out and make clear the *Categories* or fundamental modes" (Peirce 1998, 143). That Peirce is not, however, in agreement with Hegel's results, can only be inferred from Peirce's explanation of why he adopts the term: "This is the science which Hegel made his starting point, under the name of the *Phänomenologie des Geistes*,—although he considered it in a fatally narrow spirit, since he restricted himself to what actually forces itself on the mind and so colored his whole philosophy with the ignoration of the distinction of essence and existence and so gave it the nominalistic and I might say in a certain sense the pragmatoidal character in which the worst of the Hegelian errors have their origin" (Peirce 1998, 143). That Peirce here distances

our perception, action and thought—that is, of our experience. In the history of philosophy the question "what are the elements of appearance that present themselves to us every hour and every minute" (Peirce 1998, 147), has been answered in many different ways. Aristotle referred to the quality of things and categorized the types of object properties that general statements could be made about. Kant moved the categories from the object to the subject, and developed a kind of basic conceptual framework, as an orientation for the thinking of reality. Based on a critical debate with Kant and on a relational logic analysis of experience, Peirce defined a list of categories mediating between an objectivistic and subjectivistic ontology. His three categories should all be applicable to everything, and thus show only what is familiar. In the final analysis, Peirce suggested the concept of a sign as a threefold relation, in which experience and reality form a unity. This was the result of a process of scientific research, with observations ("What elements present themselves to us in every minute and every hour?") and an interpretative rule ("Our experiences can be conceived of in a relational logic") leading to the formulation of a conclusion ("There are three universal categories"). The sign concept I used thus proceeds from an argument which can be criticized, and not from a metaphysical positing.

The idea that we live in a world of signs, that we ourselves are signs and orient ourselves by interpreting of signs, provides coevolutionary science with a decisive advantage. With it, natural and mental processes can be seen as comparable, without the need to presume or postulate their identity. We can conceive of society *and* nature as sign processes,

---

himself from Hegel needs to be emphasized, especially in regards to interpretations which see Peirce in closer agreement with Hegel (cf. Habermas 1971, 142). A more complex assessment of the relationship of the two philosophers can be gained from their respective referencing of Kant: Both Hegel and Peirce observe that Kant's epistemic theory ignores the state of Being-in-the-world. Hegel starts with the observation that it is not philosophy that marks the beginning of thought but that everyone has a "faculty of thought". His objective is then to lead consciousness to what it is, and by being conscious of itself develop what everyone already latently knows and needs to know. Peirce, on the contrary, holds on to Kant's objective to explore the world as it seems to us. Being-in-the-world is not the constitutive principle of Peirce's own theory but a starting condition of scientific knowledge (see Pape 1991, 25–31).

and need not switch from one conceptual framework to the other, depending on what object we are dealing with. It remains undecided which rules and laws natural phenomena or the formations of social order will follow. Where can similarities be identified and what do basic differences consist of? Are societies emergent phenomena with their own logic, or are social relations—at least predominantly—defined by their biophysical conditions? Such are the questions for a kind of empirical research that puts up for critical discussion its explicit as well as implicit preconditions.

Sign, Purpose, and Future in the Conceptual Framework of Coevolutionary Science

But if we understand society and nature as sign processes, is this ultimately still not an extension of the sign concept developed from human practice? For here, the sign concept—developed from the specifically human, intersubjective usages of signs—is naturalized and applied to the entire cosmos. Will we thereby not lose what is specifically human (cf. Habermas 1995)? Are we perhaps simply forming an equivalent to the actor–network theory of Bruno Latour, which views all forms of practice as "exchanges between human and non-human actors" (Latour 2001, 250; translation J.J.)?

Actor–network theory does indeed lend itself to prevent misunderstandings and to distinguish the views outlined above. Developing his theory—referring to the works in natural philosophy by Isabelle Stengers and her teacher Alfred N. Whitehead—Latour fundamentally addresses the concept of nature in modernity. In his central thesis, he argues that the modern conceptual world is defined by an asymmetrical understanding of reality. Society and nature, and subject and object are differently evaluated. Each first part of the two conceptual pairs is understood as an actor and considered to have agency. Each second part is ultimately an artifact of our knowledge (cf. Latour 2004). Admittedly, modernity did never, in fact, adhere to this evaluation. "We have never been modern", says Latour (1993), for we always lived in mixed nature–society worlds. In these worlds, nature always exists as socially appropriated nature; likewise, society is always also tangible and material. But our ideas and concepts should correspond to this inter-relatedness, as well. Only then, so Latour, is there a chance for us to overcome the human-made ecological problems of our societies. He sees his actor–network theory as a theory for a new environmental politics which takes into account already in the epistemic process the coexist-ence of humans and nature.

Latour's conceptual socialization of nature may be motivated by goals we share and also deem worthwhile. But we still need to critically assess Latour's outlines of actor–network theory and learn from his failures. Latour developed his theory, going from one affront to the next. Prior to the provocative phrase that we have never been modern, and conse-quently—it must be added—live in permanent self-deceit, he developed a sociology of science that was no less provocative. In a constructivist argument, Latour questions—based on empirical studies of labora-tory research—the epistemic autonomy of the unknown and unnamed world. Provocatively he asks where lactic acid yeast was before Pasteur discovered it. His answer is even more provocative: He maintains that lactic acid yeast was nowhere before its discovery. Pasteur was the first to create a condition which allowed lactic acid to become stable. The yeast is not real in the sense that it has always been there, but it emerges as an effective variable because of Pasteur, and thus becomes what we know today. Similar to Karin Knorr-Cetina's argument about the "manufac-ture of knowledge" (Knorr-Cetina 1981), Latour implies that the world scientists explore is actually created by them.

This kind of constructivism was supposed to provoke the natural scientists in particular. And at the time it did (cf. Bricmont and Sokal 1999). Latour kept on developing his ideas in the debate with the non-polemical, factual criticism of his sociology of science. He gave his constructivism not an epistemological but an ontological bent: Everything is constructed because everything has always been con-structed, meaning "everything" not in the singular but in the plural. "Everything"—they are *actants*, writes Latour, following Algirdas Greimas. And an actant, to put it a bit disrespectfully, is an active entity, expressed within the conceptual horizon of French philosophy. Actants can be individual human beings, groups of people, or non-human things but also animals or technical artifacts—basically anything with a potential for agency. Actants have motives and put themselves in relation to other actants. Lactic acid yeast is effective in the production of wine and milk, and this is the reason—following Latour—why it is lactic acid yeast. Likewise, Pasteur, as we know him, is Pasteur because he meant to change something about the fermentation in the wine industry, or because he wanted to dissociate himself from people who claimed that spontaneous generation was to be found in something that was rotting away.

With the concept of the actant, Latour constructs a world of active entities. Actants are active because the put themselves into relation to other actants. When such networks of relations stabilize, things emerge, so Latour (2005). But he is unable to perceive and describe these things as something fixed, firm, immutable. Natural scientists feel especially provoked by Latour as his theory wholly ignores what they consider essential for their work: the resistance from the things or objects they examine and research.[59] There still has to be something else—to paraphrase their outlook—when Pasteur, for whatever reason, discards spontaneous generation as a plausible hypothesis. To put it differently, one could ask why the oppositional scientist actants of Pasteur did not create the spontaneous generation actant. "Interesting question", Latour would probably say. "Let us examine what stabilized the one network and not the other".—"No", reply the biochemists, "we have a much simpler answer".

Even without entering polemic discussions one can state that actor–network theory negates entirely the experience of resistance from the objects of research. No matter whether series of data from long

---

[59]Latour refers to "things" but he does not reproduce the classic ontological differentiation in objects and events already used in ancient Greece. A thing in actor–network theory is a stabilized actor–network where different actants are securely welded into one constellation. When we look at, for instance, the thing milk, it is not only about the white liquid but about "cows, feedstuff, milking machines, microbes, people and other components" (Kropp 2006, 224). Latour's understanding of the concept follows Whitehead's usage of the term *nexus*. But with the outline of a symmetrical ontology, Latour loses the capital which he could gain by the reception of Whitehead's metaphysics. Whitehead's theory takes into account the differences in reality (cf. Whitehead 1979, 331), which Latour simply ignores.

years of weather observations are analyzed, the occurrence of biological species and their commonness in certain regions, the function of a molecule for the growth of a plant, or the meaning of a judicial decision: Scientists always make the experience that in the pursuit of their research object, they happen upon surprising facts. They experience resistance when their research objects do not accommodate to the scientists' expectations. Actor–network theory cannot adequately address this experience. In this respect, it is stuck in the early constructivist hyperbole of Latour's sociology of science.

The ideas and thought structures, in which I embedded coevolutionary science in the previous chapters, function differently. I did not dismantle the concept of the object nor did I have to transfer the characteristic of being active—in dualistic views ascribed to the subject—onto the object. The differentiation between subject and objects is overcome with a third concept, the concept of the sign. Signs mediate between experience and reality. They represent the one in the other. Representation, understood as sign process, is synonymous with "mind". This is not about the mind psychologists deal with; "mind" here does not mean "consciousness" (cf. Peirce 1983, 169). Rather it is about the fact that any sign process is a reasoning process, not only in humans societies but also in the natural order. In the physical and biological world, we can find early forms of the intellectual capacity we see in humans, and we can, as humans, comprehend and analyze these early forms as sign processes. Mind and matter are thus a unity, as they share the same forms of development.

Other than in Latour's theory, in this version of monism that what makes for the sphere of human life will not be simply transferred onto the non-human world. Therefore purpose-oriented activity cannot become the paradigm for understanding every possible thing. All notions of panpsychism are to be avoided. There is no spirit—of whatever kind—at work in the objects, shaping them to fit its purposes. But how can we prevent circular reasoning? Any detailed logical analysis shows that teleological concepts of reality, which presuppose an inherent purpose orientation in everything that is mental and natural, move in self-fulfilling circles. Either they proclaim to know the great cosmological purpose of all processes, or they make the purely formal presumption that reality simply has to move towards a great ultimate purpose. To avoid this mistake we need a concept which can assert the internal self-determination of all—and not only human—sign processes, but without presuming a purpose. Perhaps such a concept can be developed, in reference to ideas of causalities as they were conceived of by Aristotle.

According to Aristotle, four aspects of causality have to be considered (cf. Aristotle, *Metaphysics* 1013a 24–1014a 25): Since the times of Galilei, the efficient cause (*causa efficiens*) has been attributed the central if not exclusive importance in the development of the sciences. In addition, Aristotle recognizes the material cause (*causa materialis*), the formal cause (*causa formalis*) and the final cause (*causa finalis*). To truly have knowledge about an object one needs to be able to explain all four causes. The material and formal cause define the Being of objects; the efficient and final cause define their Becoming. The final cause Aristotle takes to be the external purpose for the sake of which something takes place. Exercise exists for the sake of our health; vertebrate animals living on land have lungs so the exchange of gases in their bodies can be ensured. But aside from the external impulse—the efficient cause—we need to always take into account that each thing also exists for a particular goal, which it finally causes.

What the idea of final causation accomplishes cannot be explained solely by talking about efficient causes (cf. Pape 1989, 343–402, 1991). We can explain how rules and regularities emerge. Why is the development of a lung in vertebrates, not an isolated case? Why is a random genetic modification repeated? We need the concept of final causation to be able to describe how orders emerge from random processes and how they stabilize. The view, though, that an external purpose is at work—implying that all natural and mental processes move towards an already determined goal—is not acceptable. For then everything would be predetermined, and such an assumption is to be avoided. Instead, it is a matter of showing that processes result in a final state which is defined by a certain general characteristic. Like, for instance, the characteristic to make it possible for vertebrate animals to exchange oxygen and carbon dioxide.[60] This final state orients, but it does not determine the development process. The concept of purpose is therefore not adequate to express what final causation means. Final causes, which can be verified in physical and biological objects, are not purposes that have always been present or that a spirit put into the objects. To say that something is the result of final causation means nothing more than that it tends towards a certain result. Prior or at the beginning of the process, this result is not yet known; it emerges during the process. "Final causation" therefore means only that all sign processes tend towards final states.[61]

This view of final causation is crucial for the concept of reality in coevolutionary science, as it is developed here. On the one hand, this concept provides a foundation for the unity of the natural and social sciences/humanities (1), on the other hand it expresses, in a condensed form, the essential character of this concept of reality, temporality (2), and thus turns the focus towards the ethical commitment of coevolutionary science (3).

1. Why do we understand—despite all proneness for error—physical and biological processes just like we understand social action? An explanation can be found in the view that all our research objects move towards possible final states: Because the whole world, not only the area of human action, is defined by the pursuit of final states. Our purpose-oriented action seems to be—if you will—a

---

[60]"Final causation" can be defined as follows: "We must understand by final causation that mode of bringing facts about according to which a general description of result is made to come about. Quite irrespective of any compulsion for it to come about in this or that particular way; (...). Final causation does not determine in what particular way it is to be brought about, but only that the result shall have a certain general character" (Peirce 1931, 92/C.P. 1.211; Cf. Pape 1991, 60–66).

[61]Pape (1989, 372; translation J.J.) describes this concept of final causation like this: "A final causation is an only generally defined type of final state of a process, influencing earlier phases of this process in such a way that they show a tendency towards this final state."

specific form of the final causation that defines all objects.[62] Like Latour's actor–network theory, I thus ascertain that our understanding of the world is anthropomorphic. But contrary to Latour's conceptual world I do not sacrifice the notion of an object. Objects are, as explained above, ascribed a double function in the sign process: They can be a point of reference of sign processes, or the cause of their changes. Hence signs not only relate to their objects because of habit and routine. They function not only as symbols but also as icons and indices. Which means they can refer to an object based on analogy or based on a direct, existential relationship. The indexical relationship of sign and object, in particular, conveys the resistance of objects in our contexts of perception and communication.

2. The view that all signs are defined by final causation highlights the dimension of time inherent in the sign concept. Signs function by referring to prior objects and consequently bringing forth subsequent interpretants. Signs thus constitute the temporal order because they put events in relations of before and after. Initially, humans bartered the things they had, to get at goods they had not. Then money was invented as a medium to facilitate this exchange and to make goods more accessible for everybody. But time is not only an internal mode of all signs. We also can examine time and turn it into our research object, as all objects are constituted by signs, after all. When we regard objects as sign process, we can see that some final states are reached in sign processes, but others are not. Thus we can observe that money facilitates the access to goods of others, and yet there does not automatically appear a distributional justice of goods in the money economy. Let's look at another example: When in the course of evolution, animals moved

---

[62]"It is (...) a widespread error to think that a 'final cause' is necessarily a purpose. A purpose is merely that form of final cause which is most familiar to our experience" (Peirce 1931, 91/C.P 1.211). For Peirce, the evolution of species is an example "of a result of 'final causes' taking an actual effect in the history of nature" (Pape 1989, 373). Can a final cause be conceived of for social systems or societies, as well? It is a question demanding extensive discussion. The reader is reminded in this context of Aristotle who saw city-state moving towards self-sufficiency (Aristotle 1978, 154–157/Pol 1261b 12–13) and of Parsons (1966) who builds on this idea.

from water to land, gill breathing developed into lung breathing. But when mammals then also populated the ocean, lungs did not revert back to gills.[63] Both examples illustrate how change and development processes can be examined as the emergence of new structures. Whether emergent phenomena truly constitute something new, or whether they are derived (and thus fully explained) from preceding structures—this question is the subject of a multifaceted debate in recent ontology. I cannot here go into the details of the discussion but (independent of the outcome of the ontological debate about the assertion of the new) I see it as validation of my attempts to conceptualize time as modality.[64] Thus, the past consists of the sum of all completed, real events, whereas the future is uncertain and includes all possible events. The present is the passage from uncertainty to certainty, from future possibilities to the realness of what is past.[65] The future, however, is only in part uncertain. The shorter the distance of time to the present, the smaller the scope of possibilities and the higher the probability to correctly predict future events. The world we live in—including the social world—is structured by laws and behavioral habits. These put limitations on the range of possible future events. If we, for instance, ask a farmer to use a certain pasture for extensive instead of intensive farming, we ultimately deal with a behavioral habit. No matter how much we wish for it to be different, the farmer will only do as we ask if he gets paid for it in the near

---

[63]Natural laws describing reversible processes do not yield final states. Only irreversible processes and the laws governing them are *finious*. Peirce describes this as follows: "Those non-conservative actions which seem to violate the law of energy, and which physics explains away as due to chance action among trillions of molecules, are one and all marked by two characters. The first is that they act in one determinate direction and tend asymptotically toward bringing about an ultimate state of things. If teleological is too strong a word to apply to them, we might invent the word *finious*, to express their tendency toward a final state. The other character of non-conservative action is that they are *irreversible*" (Peirce 1958, 286f./CP 7.471).

[64]For Peirce's concept of time, cf. Peirce (1991). We can conceive of time as "the system of those relations any event has to any other event in the past, present, and future" (Peirce 1991, 482). Cf. in particular Pilot (1972, 252ff.); Pape (1989).

[65]The present then becomes a "Nascent state of the Actual" (Peirce 1934c, 313/CP 5.462), a birth state of the real.

future. In this respect the future is not open. We may find a farmer who is interested in issues of environment and nature protection, and he may as an exception pay himself for the extensification. But this does not cancel out the rule "service for money". It is still a reality just as gravity is a natural law, defining our lives. But there is a difference between the two rules with regard to the partially open future: We cannot change gravity but we can change the rule "service for money". As past events, both are immutable/unalterable. Let's say a glass fell out of our hand yesterday and dropped to the floor. We can make this fact the object of our research, asking about causes and consequences, just like the farmer's demand for money. As a future event, we may experimentally remove the effect of gravity when we e.g. go on a parabolic flight. And we can use machines, creating gravity in the weightlessness of a space flight. But gravity itself is not created by humans. We have no influence on gravity's fundamental validity; we can only influence the conditions under which it takes effect. It is different with the rule "service for money". Humans made this rule, and it is valid only as long as humans adhere to it. In principle, each present can make the decision to no longer exchange services against money but only against another service or against goods. In such decisions resides the potential for changing behavioral habits.

3. The ethical challenges for coevolutionary science lie in the partial uncertainty of the future. How should we act? What is possible? What is not possible? What development is desirable? These questions are only relevant if we presume that we can change something about our fate. If future events are predetermined, it makes no sense to think about possible changes of one's habits. The same holds true if all future events are seen as arbitrary and essentially incalculable. Hume and Kant were wrong when they thought that knowledge is not eminently important in this respect. For we can indeed correlate the purposes of our actions with what we cannot change or not yet change, the behavioral habits of human beings and the laws of nature. This does not mean that we can deduce right and wrong from the observation of society and nature. Nature and society are not statistical variables, either. But we can anticipate where these

processes of change will possibly end. Social processes are sustainable when humans adjust the purposes of their actions in a way as to not eliminate possibilities of development. To do so, we need knowledge about interrelated contexts, about self-reinforcing processes, about positive feedback and lack of feedback, and about coevolutionary dynamics.

References

Alexander, J. C. (1987). Action and Its Environments. In J. C. Alexander, B. Giesen, R. Münch, & N. J. Smelser (Eds.), *The Micro-Macro Link* (pp. 289–318). Berkeley: University of California Press.

Aristoteles. (1978). *Politik – griechisch und deutsch*, ed. F. Susemihl, Part I: Text und Übersetzung. In Aristoteles, Werke, Vol. 6. Reprint of the edition Leipzig 1879, Aalen: Scientia-Verlag.

Avital, E., & Jablonka, E. (2000). *Animal Traditions: Behavioural Inheritance in Evolution*. Cambridge: Cambridge University Press.

Beck, U. (1988). *Gegengifte. Die organisierte Unverantwortlichkeit*. Suhrkamp: Frankfurt am Main.

Beck, U. (1999). *World Risk Society*. Cambridge: Polity.

Beck, U., Giddens, A., & Lash, S. (1996). *Reflexive Modernisierung. Eine Kontroverse*. Suhrkamp: Frankfurt am Main.

Beckert, J. (1997). *Grenzen des Marktes. Die sozialen Grundlagen wirtschaftlicher Effizienz*. Frankfurt am Main and New York: Campus.

Bonner, J. T. (1980). *The Evolution of Culture in Animals*. Princeton, NJ: Princeton University Press.

Brandom, R. B. (1994). *Making It Explicit: Reasoning, Representing, and Discursive Commitment*. Cambridge, MA: Harvard University Press.

Brandom, R. B. (2001). *Begründen und Begreifen. Eine Einführung in den Inferentialismus*. Suhrkamp: Frankfurt am Main.

Bricmont, J., & Sokal, A. (1999). *Fashionable Nonsense: Postmodern Intellectuals' Abuse of Science*. New York: Picador.

Bryson, B. (2003). *A Short History of Nearly Everything*. New York: Random House.

Burger, P. (2006). *Sustainability Science: The Science of the Future*. Unpublished Manuscript.

Burkholz, R. (2008). *Problemlösende Argumentationsketten*. Weilerwist: Velbrück.

Caspar, M. (1937). Nachbericht. In J. Kepler, 1937: M. Caspar (Ed.), *Gesammelte Werke: Vol. 3. Astronomia Nova* (pp. 427–484). München: Beck.

Caspar, M. (1995). *Johannes Kepler*. Stuttgart: Kohlhammer.

Comte, A. (1903 [1844]). *A Discourse on the Positive Spirit* (E. S. Beesly, Trans.). London: William Reeves.

Connor, J. A. (2004). *Kepler's Witch. An Astronomer's Discovery of Cosmic Order Amid Religious War, Political Intrigue, and the Heresy Trial of His Mother*. San Francisco, CA: Harper.

Danneberg, L. (1988). Peirces Abduktionskonzeption als Entdeckungslogik. Eine philosophiehistorische und rezeptionskritische Untersuchung. *Archiv für Geschichte der Philosophie, 70*, 305–326.

Danneberg, L. (1989). *Methodologien: Struktur, Aufbau und Evaluation*. Berlin: Duncker & Humblot.

Dennett, D. C. (1991). *Consciousness Explained*. Boston: Little, Brown and Co.

Durkheim, E. (1995 [1895]). *Die Regeln der soziologischen Methode* (R. König, Ed. and Intro.). Frankfurt am Main: Suhrkamp.

Durkheim, E. (1996 [1893]). *Über soziale Arbeitsteilung. Studie über die Organisation höherer Gesellschaften*. Frankfurt am Main: Suhrkamp.

Eco, U. (1983). Horns, Hooves, Insteps: Some Hypotheses on Three Types of Abduction. In U. Eco & T. A. Sebeok (Eds.), *The Sign of Three—Dupin, Holmes, Peirce* (pp. 198–220). Bloomington: Indiana University Press.

Eder, K. (1976). *Die Entstehung staatlich organisierter Gesellschaften. Ein Beitrag zu einer Theorie sozialer Evolution*. Suhrkamp: Frankfurt am Main.

Elias, N. (1970). *Was ist Soziologie?* München: Juventa [also published in English: Elias, 1978: What Is Sociology? London: Hutchinson].

Elias, N. (1998 [1939]). *Über den Prozeß der Zivilisation* (2., rev. and expanded ed.). Frankfurt am Main: Suhrkamp [also published in English: Elias. (2012). *On the Process of Civilisation*, ed. S. Mennell, E. Dunning, J. Goudsblom, & R. Kilminster. Dublin: UCD Press].

Emirbayer, M., & Mische, A. (1998). What Is Agency? *American Journal of Sociology, 103*(4), 962–1023.

Esser, H. (1996). *Soziologie. Allgemeine Grundlagen*. Frankfurt am Main and New York: Campus.

Fischer, H., Kumke, T., Lohmann, G., Flöser, G., Miller, H., von Storch, H., & Negendank, J. F. W. (Eds.). (2004). The Climate in Historical Times. Towards a Synthesis of Holocene Proxy Data and Climate Models. In *Proceedings of the Third GKSS School on Environmental Research*. Berlin: Springer.

Foukal, P., Fröhlich, C., Spruit, H., & Wigley, T. M. L. (2006). Variations in Solar Luminosity and Their Effect on the Earth's Climate. *Nature, 443*, 161–166.

Giesen, B. (1999). Codes kollektiver Identität. In W. Gephart & H. Waldenfels (Eds.), *Religion und Identität. Im Horizont des Pluralismus* (pp. 13–43). Frankfurt am Main: Suhrkamp.

Gigerenzer, G. (1991). From Tools to Theories: A Heuristic of Discovery in Cognitive Psychology. *Psychological Review, 98*(2), 254–267.

Gilder, J., Gilder, A.-L. (2004). *Heavenly Intrigue. Johannes Kepler, Tycho Brahe, and the Murder Behind One of History's Greatest Scientific Discoveries.* New York: Doubleday.

Grunenberg, H. (2005). Rezension zu: Jo Reichertz (2003). Die Abduktion in der qualitativen Sozialforschung. *Forum Qualitative Sozialforschung/ Forum: Qualitative Social Resarch, 6*(2), Art. 17. Retrieved August 2, 2005.http://www.qualitativeresearch.net/fqs-texte/2-05/05-2-17-d.htm.

Habermas, J. (1971). *Erkenntnis und Interesse.* Frankfurt am Main: Suhrkamp.

Habermas, J. (1976). *Zur Rekonstruktion des Historischen Materialismus.* Frankfurt am Main: Suhrkamp.

Habermas, J. (1981a). *Theorie des kommunikativen Handelns. Band 1: Handlungsrationalität und gesellschaftliche Rationalisierung.* Frankfurt am Main: Suhrkamp [also published in English: Habermas, J. (1984). *Theory of Communicative Action, Volume One: Reason and the Rationalization of Society* (T. A. McCarthy, Trans.). Boston, MA: Beacon Press].

Habermas, J. (1981b). *Theorie des kommunikativen Handelns. Band 2: Zur Kritik der funktionalistischen Vernunft.* Frankfurt am Main: Suhrkamp [also published in English: Habermas, J. (1987). *Theory of Communicative Action, Volume Two: Lifeworld and System: A Critique of Functionalist Reason* (T. A. McCarthy, Trans.). Boston, MA: Beacon Press].

Habermas, J. (1995). Peirce and Communication. In K. L. Ketner (Ed.), *Peirce and Contemporary Thought. Philosophical Inquiries* (pp. 243–266). New York: Fordham University Press.

Habermas, J. (1999). *Wahrheit und Rechtfertigung. Philosophische Aufsätze.* Frankfurt am Main: Suhrkamp [also published in English: Habermas. (2003). *Truth and Justification* (B. Fultner, Ed. and Trans.). Cambridge: Polity Press].

Hanson, N. R. (1965 [1958]). *Patterns of Discovery: An Inquiry into the Conceptual Foundations of Science.* London: Cambridge University Press.

Hawking, S. (2000). *Die illustrierte kurze Geschichte der Zeit.* Reinbek bei Hamburg: Rowohlt.

IPCC. (1996). Climate Change 1995: The Science of Climate Change. In J. T. Houghton, L. G. M. Filho, B. A. Callander, N. Harris, A. Kattenberg, & K. Maskell (Eds.), *Contribution of Working Group I to the Second Assessment Report of the Intergovernmental Panel on Climate Change*. New York: Cambridge University Press.

IPCC. (2001). Climate Change 2001: The Scientific Basis. In J. T. Houghton, Y. Ding, D. J. Griggs, M. Noguer, P. J. van der Linden, X. Dai, K. Maskell, & C. A. Johnson (Eds.), *Contribution of Working Group I to the Third Assessment Report of the Intergovernmental Panel on Climate Change*. New York: Cambridge University Press.

IPCC. (2007). Climate Change 2007: The Physical Science Basis. In S. Solomon, D. Qin, M. Manning, Z. Chen, M. Marquis, K. B. Averyt, M. M. B. Tignor, & H. L. Miller, Jr. (Eds.), *Contribution of Working Group I to the Fourth Assessment Report of the Intergovernmental Panel on Climate Change*. New York: Cambridge University Press.

Jetzkowitz, J. (1996). *Störungen im Gleichgewicht. Das Problem des sozialen Wandels in funktionalistischen Handlungstheorien* [Marburger Beiträge zur Sozialwissenschaftlichen Forschung; Bd. 7], Münster: LIT-Verlag.

Jetzkowitz, J. (2010). "Menschheit", "Sozialität" und "Gesellschaft" als Dimensionen der Soziologie. Anregungen aus der Nachhaltigkeitsforschung. In G. Albert, R. Greshoff, & R. Schützeichel (Eds.), *Dimensionen und Konzeptionen von Sozialität*. Wiesbaden: VS Verlag für Sozialwissenschaften, 257–268.

Joas, H. (1992). *Die Kreativität des Handelns*. Frankfurt am Main: Suhrkamp.

Jobe, T. H. (1986). Review of "Occult and Scientific Mentalities in the Renaissance" by Brian Vickers. *The Sixteenth Century Journal, XVII*(1), 113–115.

Kamper, D. (1997). Mensch. In C. Wulff (Ed.), *Vom Menschen. Handbuch Historische Anthropologie* (pp. 85–91). Weinheim und Basel: Beltz.

Kant, I. (1899 [1781/1787]). *The Critique of Pure Reason* (J. M. D. Meiklejohn, Trans.). New York, NY: Willey Book Co.

Kant, I. (1999). *Correspondence* (A. Zweig, Trans.). Cambridge: Cambridge University Press.

Kapitan, T. (1994). Inwiefern sind abduktive Schlüsse kreativ? In H. Pape (Ed.), *Kreativität und Logik: Charles S. Peirce und das philosophische Problem des Neuen* (pp. 144–158). Frankfurt am Main: Suhrkamp.

Keil, G. (1985). *Philosophiegeschichte I. Von der Antike bis zur Renaissance*. Stuttgart: Kohlhammer.

Kelle, U. (1994). *Empirisch begründete Theoriebildung. Zur Logik und Methodologie interpretativer Sozialforschung.* Weinheim: Deutscher Studien Verlag.

Knorr-Cetina, K. (1981). *The Manufacture of Knowledge: An Essay on the Constructivist and Contextual Nature of Science.* Oxford and New York: Pergamon Press.

Koestler, A. (1959). *The Sleepwalkers. A History of Man's Changing Vision of the Universe.* Macmillan: New York.

König, R. (1967). Biosoziologie. In R. König (Ed.), *soziologie* (pp. 48–53). Frankfurt am Main: Fischer.

Kovarik, W. (2004). *Environmental History Timeline.* Retrieved November 10, 2004. http://www.radford.edu/~wkovarik/envhist/.

Krafft, F. (1973). Johannes Keplers Beitrag zur Himmelsphysik. In F. Krafft, K. Meyer, & B. Sticker (Eds.), *Internationales Kepler Symposium, Weil der Stadt, 1971, Referate und Diskussionen* (pp. 55–139). Hildesheim: Gerstenberg.

Krafft, F. (2005). Johannes Kepler – Die neue, ursächlich begründete Astronomie. In J. Kepler *Astronomia Nova. Neue, ursächlich begründete Astronomie* (F. Krafft, Trans.). Wiesbaden: Marix (pp. V–LVIV).

Kropp, C. (2006). "Enacting Milk": Die Akteur-Netz-Werke von "Bio-Milch". In M. Voss & B. Peuker (Eds.), *Verschwindet die Natur? Die Akteur-Netzwerk-Theorie in der umweltsoziologischen Diskussion* (pp. 203–232). Bielefeld: transcript.

Kuhn, T. S. (1962). *The Structure of Scientific Revolutions.* Chicago: University of Chicago Press.

Kuhn, T. S. (1977). Neue Überlegungen zum Begriff des Paradigma. In T. S. Kuhn, *Die Entstehung des Neuen. Studien zur Struktur der Wissenschaftsgeschichte* (389–420). Frankfurt am Main: Suhrkamp.

Latour, B. (1993 [1991]). *We Have Never Been Modern* (C. Porter, Trans.). Cambridge, MA: Harvard University Press.

Latour, B. (2001). Eine Soziologie ohne Objekt? Anmerkungen zur Interobjektivität. *Berliner Journal für Soziologie, 11*(2), 237–252.

Latour, B. (2004). *Politics of Nature: How to Bring the Sciences into Democracy* (C. Porter, Trans.). Cambridge, MA: Harvard University Press.

Latour, B. (2005). *Reassembling the Social: An Introduction to Actor-Network-Theory.* Oxford: Oxford University Press.

Latour, B., & Woolgar, S. (1979). *Laboratory Life. The Social Construction of Scientific Facts.* Beverly Hills: Sage.

Lemcke, M. (2002). *Johannes Kepler.* Reinbek bei Hamburg: Rowohlt.

Lipton, P. (1991). *Inference to the Best Explanation*. London: Routledge.

Luhmann, N. (1984). *Soziale Systeme. Grundriß einer allgemeinen Theorie*. Frankfurt am Main: Suhrkamp [also published in English: Luhmann. (1995). *Social Systems*. Stanford: Stanford University Press].

Luhmann, N. (1988). *Die Wirtschaft der Gesellschaft*. Frankfurt am Main: Suhrkamp.

Luhmann, N. (1989a). *Ecological communication* (J. Bednarz, Trans.). Chicago: University of Chicago Press [originally published in German: Luhmann. (1986). *Ökologische Kommunikation. Kann die moderne Gesellschaft sich auf ökologische Gefährdungen einstellen?* Opladen: Westdeutscher Verlag].

Luhmann, N. (1989b). Politische Steuerung. Ein Diskussionsbeitrag. *Politische Vierteljahresschrift, 30*(1), 4–9.

Luhmann, N. (1990). *Die Wissenschaft der Gesellschaft*. Frankfurt am Main: Suhrkamp.

Luhmann, N. (1992). Wer kennt Wil Martens? *Kölner Zeitschrift für Soziologie und Sozialpsychologie, 44*(1), 139–142.

Luhmann, N. (1997). *Die Gesellschaft der Gesellschaft*. Frankfurt am Main: Suhrkamp. [also published in English: Luhmann. (2012/2013). *Theory of Society*. Stanford: Stanford University Press].

Luhmann, N. (2002). *Die Politik der Gesellschaft* (A. Kieserling). Frankfurt am Main: Suhrkamp.

Matthes, J. (1985). Die Soziologen und ihre Wirklichkeit. In W. Bonß & H. Hartmann (Eds.), *Entzauberte Wissenschaft. Sonderband 3 der Sozialen Welt* (pp. 49–64). Göttingen: Otto Schwartz und Co.

Maus, H. (1967). Zur Vorgeschichte der empirischen Sozialforschung. In R. König (Ed.), *Handbuch der empirischen Sozialforschung* (Vol. 1, pp. 21–56). Stuttgart: Ferdinand Enke.

Mayntz, R. (Ed.). (1980). *Implementation politischer Programme. Empirische Forschungsberichte*. Athenäum: Königstein, Ts.

Mayntz, R. (Ed.). (1983). *Implementation politischer Programme II – Ansätze zur Theoriebildung*. Opladen: Westdeutscher Verlag.

Mayntz, R. (1987). Politische Steuerung und gesellschaftliche Steuerungsprobleme. Anmerkungen zu einem theoretischen Paradigma. In T. Ellwein, J. J. Hesse, R. Mayntz, & F. W. Scharpf (Eds.), *Jahrbuch zur Staats- und Verwaltungswissenschaft* (Vol. 1, pp. 89–109). Baden-Baden: Nomos.

Mayntz, R., & Scharpf, F. W. (Eds.). (1995). *Gesellschaftliche Selbstregulierung und politische Steuerung*. Frankfurt am Main: Campus.

Mead, G. H. (1967). *Mind, Self, and Society from the Standpoint of a Social Behaviorist* (C. W. Morris, Ed., with introduction). Chicago and London: The University of Chicago Press.

Meléghy, T. (2003). Methodologische Grundlagen einer evolutionären Soziologie. In T. Meléghy & H.-J. Niedenzu (Eds.), *Soziale Evolution. Die Evolutionstheorie und die Sozialwissenschaften. Sonderband 7 der Österreichischen Zeitschrift für Soziologie* (pp. 114–146). Opladen: Westdeutscher Verlag.

Merton, R. K. (1936). The Unanticipated Consequences of Purposive Social Action. *American Sociological Review, 1*(6), 894–904.

Mittelstraß, J., Schroeder-Heister, P. (1997). Zeichen, Kalkül, Wahrscheinlichkeit. Elemente einer Mathesis universalis bei Leibniz. In H. Stachowiak (Ed.), *Pragmatik. Handbuch Pragmatisches Denken* (Vol. 1, pp. 392–414). Darmstadt: Wissenschaftliche Buchgesellschaft.

Münch, R., & Lahusen, C. (2001). *Democracy at Work: A Comparative Sociology of Environmental Regulation in the United Kingdom, France, Germany, and the United States.* Westport, CT: Praeger.

Oehler, K. (1995). A Response to Habermas. In K. L. Ketner (Ed.), *Peirce and Contemporary Thought. Philosophical Inquiries* (pp. 267–271). New York: Fordham University Press.

Oevermann, U. (1991). Genetischer Strukturalismus und das sozialwissenschaftliche Problem der Erklärung der Entstehung des Neuen. In S. Müller-Doohm (Ed.), *Jenseits der Utopie. Theoriekritik der Gegenwart* (pp. 267–336). Frankfurt am Main: Suhrkamp.

Ott, K., & Döring, R. (2008). *Theorie und Praxis starker Nachhaltigkeit.* Marburg: Metropolis-Verlag.

Pape, H. (1989). *Erfahrung und Wirklichkeit als Zeichenprozeß. Charles S. Peirces Entwurf einer Spekulativen Grammatik des Seins.* Suhrkamp: Frankfurt am Main.

Pape, H. (1991). Einleitung. In C. S. Peirce, *Naturordnung und Zeichenprozeß. Schriften über Semiotik und Naturphilosophie* (pp. 11–109). Frankfurt am Main: Suhrkamp.

Pape, H. (1994). Zur Einführung: Logische und metaphysische Aspekte einer Philosophie der Kreativität. C. S. Peirce als Beispiel. In H. Pape (Ed.), *Kreativität und Logik: Charles S. Peirce und das philosophische Problem des Neuen* (pp. 9–59). Frankfurt am Main: Suhrkamp.

Pape, H. (1995). *The Social Nature of Reality and Communication: Peirce vs. Mead?* Unpublished Manuscript.

Pape, H. (1999). Abduction and the Typology of Human Cognition. In: *Transactions of the Charles S. Peirce Society, 35*(2), 248–269.

Pape, H. (2002). Indexikalität und die Anwesenheit der Welt in der Sprache. In M. Kettner & H. Pape (Eds.), *Indexikalität und sprachlicher Weltbezug* (pp. 91–119). Paderborn: Mentis.

Parsons, T. (1964). A Functional Theory of Change. In A. Etzioni & E. Etzioni (Eds.), *Social Change: Sources, Patterns, and Consequences* (pp. 83–97). New York: Basic Books.

Parsons, T. (1966). *Societies: Evolutionary and Comparative Perspectives*. Englewood Cliffs, NJ: Prentice Hall.

Parsons, T. (1971). *The System of Modern Societies*. Englewood Cliffs, NJ: Prentice Hall.

Parsons, T. (1977). Comparative Studies and Evolutionary Change. In *Parsons, Talcott, Social Systems and the Evolution of Action Theory* (pp. 279–320). New York and London: The Free Press.

Peirce, C. S. (1931). The Classification of Sciences. In, C. Hartshorne & P. Weiss (Eds.), *Collected Papers, Vol. I: Principles of Philosophy* (pp. 75–137). Cambridge, MA: Harvard University Press.

Peirce, C. S. (1934a). Lectures on Pragmatism. In C. Hartshorne & P. Weiss (Eds.), *Collected Papers, Vol. V: Pragmatism and Pragmaticism* (pp. 13–131). Cambridge, MA: Harvard University Press.

Peirce, C. S. (1934b). Some Consequences of Four Incapabilities. In C. Hartshorne & P. Weiss (Eds.), *Collected Papers, Vol. V: Pragmatism and Pragmaticism* (pp. 156–189). Cambridge, MA: Harvard University Press.

Peirce, C. S. (1934c). The Fixation of Belief. In C. Hartshorne & P. Weiss (Eds.), *Collected Papers, Vol. V: Pragmatism and Pragmaticism* (pp. 293–313). Cambridge, MA: Harvard University Press.

Peirce, C. S. (1958). *Collected Papers, Vol. VII: Science and Philosophy* (A. W. Burks, Ed.). Cambridge, MA: Harvard University Press.

Peirce, C. S. (1976). *The New Elements of Mathematics, Vol 4: Mathematical Philosophy* (C. Eisele, Ed.). The Hague: Mouton Publishers.

Peirce, C. S. (1983 [1903]). *Phänomen und Logik der Zeichen* (H. Pape, Ed. and Trans.). Frankfurt am Main: Suhrkamp.

Peirce, C. S. (1986 [1901]). Minutiöse Logik. Aus den Entwürfen zu einer Logik. In C. S. Peirce, *Semiotische Schriften* (Vol. I, Ed. and Trans. C. Kloesel & H. Pape, pp. 376–408). Frankfurt am Main: Suhrkamp.

Peirce, C. S. (1991). *Naturordnung und Zeichenprozeß. Schriften über Semiotik und Naturphilosophie* (H. PapeEd. and Intro.). Frankfurt am Main: Suhrkamp.

Peirce, C. S. (1998). *The Essential Peirce: Selected Philosophical Writings, Vol. 2 (1893–1913)* (N. Houser et al., Ed.). Bloomington: Indiana University Press.

Pilot, H. (1972). *Prolegomena zu einer kritischen Theorie der Erfahrung*. Heidelberg: Dissertation Philosophisch-historische Fakultät.

Platon. (1990a). Menon. In G. Eigler (Ed.), *Platon, Werke in acht Bänden. Griechisch – Deutsch* (Vol. 2, pp. 505–599). Darmstadt: Wissenschaftliche Buchgesellschaft.

Platon. (1990b). Politeia. In G. Eigler (Ed.), *Platon, Werke in acht Bänden. Griechisch – Deutsch* (Vol. 4). Darmstadt: Wissenschaftliche Buchgesellschaft.

Popper, K. R. (1950 [1945]). *The Open Society and Its Enemies.* Princeton, NJ: Princeton University Press.

Popper, K. R. (1963). *Conjectures and Refutations: The Growth of Scientific Knowledge.* London: Routledge and Kegan Paul.

Popper, K. R. (1994 [1934]). Logik der Forschung. Tübingen: Mohr [also published in English: Popper. (1965). *The Logic of Scientific Discovery.* New York: Harper & Row].

Portmann, A. (1972). Tiersoziologie. In W. Bernsdorf (Ed.), *Wörterbuch der Soziologie* (3rd ed., pp. 857–861). Frankfurt am Main: Fischer.

Prigogine, I. (1980). *From Being to Becoming: Time and Complexity in the Physical Sciences.* San Francisco: Freeman.

Prigogine, I., & Stengers, I. (1984). *Order Out of Chaos: Man's New Dialogue with Nature.* New York: Bantam Books.

Radkau, J. (2008). *Nature and Power: A Global History of the Environment* (T. Dunlap, Trans.). Cambridge: Cambridge University Press.

Raeithel, A. (1994). Symbolic Production of Social Coherence. The Evolution of Dramatic, Discursive and Objectified Meaning Systems. *Mind, Culture and Activity, 1*(1–2), 69–123.

Rammert, W. (1997). New Rules of Sociological Method: Rethinking Technology Studies. *British Journal of Sociology, 48*(2), 171–191.

Reichenbach, H. (1938). *Experience and Prediction: An Analysis of the Foundations and the Structure of Knowledge.* Chicago: University of Chicago Press.

Reichertz, J. (1991). *Aufklärungsarbeit. Kriminalpolizisten und Feldforscher bei der Arbeit.* Stuttgart: Ferdinand Enke.

Reichertz, J. (2003). *Die Abduktion in der qualitativen Sozialforschung.* Opladen: Leske+Budrich.

Ridley, M. (2003). *Nature via Nurture: Genes, Experience, and What Makes Us Human.* New York: Harper Collins.

Rogers, E. M. (1962). *Diffusion of Innovations.* New York: The Free Press of Glencoe.

Scharpf, F. W. (1989). Politische Steuerung und politische Institutionen. *Politische Vierteljahresschrift, 30*(1), 10–21.

Schröer, H. (1990). Art. Kybernetik. *Theologische Realenzyklopädie, 20*, 356–359.

Schumpeter, J. A. (1912). *Theorie der wirtschaftlichen Entwicklung.* Leipzig: Duncker & Humblot.

Schwinn, T. (2003). Makrosoziologie jenseits von Gesellschaftstheorie. Funktionalismuskritik nach Max Weber. In J. Jetzkowitz & C. Stark (Eds.), *Soziologischer Funktionalismus. Zur Methodologie einer Theorietradition* (pp. 83–109). Opladen: Leske+Budrich.

Searle, J. R. (1983). *Intentionality: An Essay in the Philosophy of Mind.* Cambridge: Cambridge University Press.

Sellars, W. (1997). *Empiricism and the Philosophy of Mind* (R. Rorty, With an introduction, a study guide by R. R. Brandom). Cambridge, MA: Harvard University Press.

Simon, H. A. (1973). Does Scientific Discovery Have a Logic? *Philosophy of Science, 40*, 471–480.

Smith, T. M., & Reynolds, R. W. (2005). A Global Merged Land-Air-Sea Surface Temperature Reconstruction Based on Historical Observations (1880–1997). *Journal of Climate, 18*(12), 2021–2036.

Stark, C. (1998). *Die blockierte Demokratie. Kulturelle Grenzen der Politik im deutschen Immissionsschutz.* Baden-Baden: Nomos.

Stark, C. (2003). Neopositivistische Gesellschaftstheorie. Ein Essay vom 'Ende der Geschichte' und zur 'natürlichen Ordnung' des Funktionalismus. In J. Jetzkowitz & C. Stark (Eds.), *Soziologischer Funktionalismus. Zur Methodologie einer Theorietradition* (pp. 219–246). Opladen: Leske+Budrich.

Stehr, N., & von Storch, H. (1999). *Klima – Wetter – Mensch.* München: Beck.

Stephenson, B. (1987). *Kepler's Physical Astronomy.* New York: Springer.

Tembrock, G. (1997). Grundlagen und Probleme einer allgemeinen Tiersoziologie. *Ethik und Sozialwissenschaften, 8*(1), 71–80.

Tenbruck, F. H. (1981). Emile Durkheim oder die Geburt der Gesellschaft aus dem Geist der Soziologie. *Zeitschrift für Soziologie, 10*(4), 333–350.

Tenbruck, F. H. (1989). Gesellschaftsgeschichte oder Weltgeschichte? *Kölner Zeitschrift für Soziologie und Sozialpsychologie, 41*(3), 417–439.

Thagard, P. R. (1988). *Computational Philosophy of Science.* Cambridge, MA and London: The MIT Press.

Tiner, J. H. (1977). *Johannes Kepler. Giant of Faith and Science.* Milford, MI: Mott Media.

Touraine, A. (1986). Krise und Wandel des sozialen Denkens. In J. Berger (Ed.), *Die Moderne – Kontinuitäten und Zäsuren* (pp. 15–39). Göttingen: Otto Schwartz und Co.

Tovey, H. (2003). Theorising Nature and Society in Sociology: The Invisibility of Animals. *Sociologia Ruralis, 43*(3), 196–215.

van Orman Quine, W. (1994). *From Stimulus to Science.* Cambridge, MA: Harvard University Press.

Vickers, B. (Ed.). (1984). *Occult and Scientific Mentalities in the Renaissance.* London: Cambridge University Press.

Viehöver, W. (2003). Die Klimakatastrophe als ein Mythos der reflexiven Moderne. In L. Clausen, E. M. Geenen, & E. Macamo (Eds.), *Entsetzliche soziale Prozesse. Theorie und Empirie der Katastrophen* (pp. 247–286). Münster: Lit.

Voelkel, J. R. (2001). *The Composition of Kepler's Astronomia Nova.* Princeton and Oxford: Princeton University Press.

Von Weizsäcker, C. F. (1971). *Die Einheit der Natur. Studien.* München: Carl Hanser [also published in English: Weizsäcker. (1980). *The Unity of Nature* (F. J. Zucker, Trans.). New York: Farrar, Straus, Giroux].

Wals, A. E. J. (2007). *Social Learning Towards a Sustainable World: Principles, Perspectives, and Praxis.* Wageningen: Wageningen Academic Publishers.

Whitehead, A. N. (1925). *Science and the Modern World.* New York: Macmillan.

Whitehead, A. N. (1979 [1929]). *Prozeß und Realität. Entwurf einer Kosmologie.* Übers. u. mit einem Nachwort v. Hans Günter Holl. Frankfurt am Main: Suhrkamp.

Witzany, G. (2000). *Life: The Communicative Structure.* Norderstedt: Libri Books on Demand.

Zachos, J., Pagani, M., Sloan, L., Thomas, E., & Billups, K. (2001). Trends, Rhythms, and Aberrations in Global Climate 65 Ma to Present. *Science, 292*(5517), 686–693.

Zilsel, E. (1976). *Die sozialen Ursprünge der neuzeitlichen Wissenschaft. Edited by Wolfgang Krohn.* Frankfurt am Main: Suhrkamp.

Zimmerli, W. C. (1997). Zeit als Zukunft. In A. Gimmler, M. Sandbothe, & W. C. Zimmerli (Eds.), *Die Wiederentdeckung der Zeit. Reflexionen, Analysen, Konzepte* (pp. 126–147). Darmstadt: Wissenschaftliche Buchgesellschaft.

4

Perspectives of Coevolutionary Science in Sustainability Discourse

Challenges of the Present—And How Coevolutionary Science Could Help Mastering Them

We are part of one society, everywhere and at all times. Today, this society is globally connected, and people living in these globally linked societies enjoy a level of material wealth that is, arguably, historically unprecedented (cf. e.g. Clark 2007, 1–4). However, this wealth is also distributed extremely unevenly. It was (and continues to be) generated primarily in societies of the so-called Global North where particular sociocultural, economic, political, and last but not least technological structures evolved. The exploitation of cheap labor and cheap natural resources mainly in the societies of the Global South is an essential determinant of this success story (cf. Williams 1944; Wallerstein 1974–1989; Moore 2015; Brand and Wissen 2017; Lessenich 2017).

It is not a new insight that the development models of the wealthy societies of the Global North, and the externalization of socio-ecological costs (cf. Lessenich 2017) that comes with them, are not sustainable. In fact, this has been acknowledged by large parts of the global public.

© The Author(s) 2019
J. Jetzkowitz, *Co-Evolution of Nature and Society*,
https://doi.org/10.1007/978-3-319-96652-6_4

However, the conclusions and practical political consequences that are drawn from this insight differ widely.

Since about 2015, the international political community has been searching systematically for development paths that offer viable alternatives to the "imperial way of life" (Brand and Wissen 2017) of the societies of the Global North. The 2030 Agenda with its list of Sustainable Development Goals (SDGs), the Paris Agreement as well as the Sendai Framework for Disaster Risk Reduction provide reference points for this search. In particular, the SDGs have moved into the center of international policy efforts as a connective framework. The 17 goals and 169 targets take into account both socioeconomic as well as ecological aspects of global development. They have effectively created a comprehensive, holistic orientation framework within which it is possible to search for different ways of creating more sustainable global structures (and which does not commit the so-called developing countries to the goals of the industrialized societies, as was implicitly demanded in the context of the Millennium Development Goals) (cf. e.g. Easterly 2006; Fukuda-Parr 2017).

The multiplicity of goals does, however, create ambivalences. Knowledge about interdependencies—i.e. knowledge that is generated from a coevolutionary science perspective—is indispensable for exploring and pursuing more sustainable developments paths. This can easily be demonstrated when looking at various conflicts of goals. Focusing on the fight against poverty and hunger (SDGs 1 and 2) and on sustainable economic growth and decent work (SDG 8) may lead to a situation where the conservation of life underwater and on land (SDGs 14 and 15) is neglected or, respectively, where the latter two goals are given lower priority.

Conversely, the diversity of the SDGs and their inherent ambivalences and potential conflicts may also create productive disturbances. Focusing on one goal and refusing to be distracted from it may very well be the most efficient path for reaching this goal—but only this *one* goal, after all. The consequence of such an approach, as Richard Norgaard (1994) has already demonstrated some time ago, is unsustainability. Precisely because the SDGs define different, in part conflicting goals, it is necessary to leave the beaten track and open up to new ways of thinking. This increases the chances that new analysis and governance strategies are being developed and implemented, strategies that can more adequately deal with the complex internal dynamics and interactions between nature and human society.

At the same time, it is necessary to pay attention to the uncertainties that ambivalences produce, and the concomitant potential for instability. Periods of instability may indeed lead to the reconfiguration of a structure. But the longer such periods last, the more likely it becomes that old wine is simply filled into new bottles, in order to not endanger the survival of the entire system. It is difficult to say whether the timeframe of 15 years that the UN experts have allowed for reconfigurations in the 2030 Agenda, should already be considered too long a period of time. Within the scientific community, at least, processes have been launched to dissolve ambivalences and establish clarity, as much as can be told by tracing publications that analyze synergies and tradeoffs between various target achievement processes (cf. e.g. Pradhan et al. 2017; Nilsson et al. 2016; Le Blanc 2015). In other words, the race for who will determine the development paths of the future is well underway.

In order to study systematically which structures and processes influencing each other with what kind of resulting interrelations, it is of crucial importance that nature and society are equally taken into consideration.[1] Thus any approach, whether from the natural or the social sciences, should be based on the current state of debate within the discipline. Furthermore, neither approach must view other aspects than their own as residual categories or, worse, depict them as negligible factors. This applies to the social sciences in particular, for their subject areas are in large parts screened and selected by the interests of established power structures. As a consequence, social science research tends to focus on individualizable aspects of behavior rather than aspects that

---

[1]This is also confirmed by the various approaches present in sustainability discourse, for exploring the interactions between nature and society. Cf. Haraway (1985), Norgaard (1994), Fischer-Kowalski and Weisz (1999), Latour (2004), Mol and Spaargaren (2006), and Ostrom (2009).

are structural–relational[2]—e.g. attitudes that can be changed, or types of behavior that can be controlled or choices that can be influenced. At most, this kind of research will consider that education and learning processes need to be initiated to reach certain goals. Invariably the focus is on reaching goals that were formulated without any (or only indirect) input of the individuals concerned (or even the general public). Little attention so far has been paid to how social processes take on a life of their own and what are the consequences of such independent developments. Cases in point are milieus that close themselves off from one other, or organizations which while working toward their official goals mostly seek to secure their continued existence.[3]

Knowledge about coevolutions clearly constitutes a reliable basis to search for sustainable development paths in globally connected societies. But from the perspective of coevolutionary science, it needs to be emphasized that such paths of development can only be found systematically if social phenomena are integrated into the study of interrelations in non-reductive ways. Which aspects need to be considered I discuss by way of example of those strategies that are currently used in sustainability discourse to produce knowledge about the interactions between nature and society.

Methods and Strategies and How They Can Be Improved

To increase knowledge about coevolutions, two methods are often used: indicator analysis and scenario analysis relying on mathematical models. Both methods come with advantages as well as drawbacks.

---

[2]Cf. Shove (2010, 1283). She trenchantly points out: "It is … clear that policy makers are highly selective in the models of change on which they draw, and that their tastes in social theory are anything but random. An emphasis on individual choice has significant political advantages and in this context [of climate change politics, J. J.], to probe further, to ask how options are structured, or to inquire into the ways in which governments maintain infrastructures and economic institutions, is perhaps too challenging to be useful".

[3]A promising exception is the Multilevel Perspective (MLP). Cf. Geels (2010, 2012a, b) and Geels and Schot (2007).

Sustainability researchers value them for a variety of reasons: first, because with them reliable statements about different aspects of reality can be made; second, because they—in different ways—offer possibilities to correlate different aspects of reality; third, because both strategies can be combined. By studying the interactions between natural and social phenomena on the basis of indicators through modeling and simulation studies, this research produces something that is highly appreciated by political decision-makers, namely statements about the future development potentials of socio-ecological systems.

Looking at this "joint venture" of indicators and modelings from a social science perspective, it seems apposite to add several remarks and considerations. They all focus on the criticism outlined above, of the limited view of social aspects. Social, economic, and demographic indicators measure conditions and represent social developments over time. They do not account for social differentiations, whether class structures, or those of social strata or milieus, nor do they consider sociocultural particularities or the specialization of social subsystems.[4] Yet, humans live and act through and within those structures.[5] This is why the question needs to be addressed to what extent the indicators of the modeling process have been adequately considered in terms of their own dependence on sociocultural factors. And in general, a coevolutionary approach will have to ask in which contexts the systematic coupling of model development and empirical research of natural and social structures are necessary and whether it should be advanced further.[6]

With these general remarks in mind, it is instructive to look in greater depth at indicators as well as at models, simulations, and scenarios—to be able to better assess the advantages and disadvantages of these methods for sustainability research and sustainability policies.

---

[4]Exceptions are models that distinguish, for example, between different actors on the basis of their lifestyles.

[5]In this context, the above-mentioned Thomas theorem should be recalled again. Cf. p. 32.

[6]Cf. e.g. Kizuka et al. (2014) and Compton et al. (2012).

Indicators

The word "indicator" is derived from the Latin word *indicare*, meaning "to point out, to inform". The person who observes and analyzes an indicator is informed by it about a fact or circumstance related to it.[7] Depending on whether the indicator is based on a simple or complicated measuring process, the information it transmits about the related situation, is, in turn, more or less dependent on preceding processes of perception and evaluation. To be able to use the distribution and number of a particular fish, say the river trout, as an indicator for the quality of flowing waters, specialized knowledge is needed to be able to recognize specific signs as indicators. Complex bio-indicators—such as lichen exposure, which is used to ascertain degrees of air pollution—are determined by complicated processes. The same is true for social and economic indicators that are used to assess living conditions and their subjective reception.

Whether simple or complex, the analysis of any given indicator requires knowledge about complex facts of nature and society. To be sure, indicators point to causes. If the oxygen levels in a flowing water decrease and pollution increases, river trout will not survive in this particular body of water. However, the causes for reduced oxygen levels and increased pollution cannot be explained by pointing to the indicator. When connections are simple or conditions clearly apparent, the causes may be obvious, but in more complex situations they are not. And this is true not only for bio-indicators but especially for indicators that point to social or economic conditions.

This needs to be taken into account especially in situations where the values of indicators change. The indicator itself cannot explain what mechanisms are causing the changes. Whoever uses changes of complex indicators as an opportunity to speculate about the causes, plainly risks to arrive at false conclusions.

Furthermore, it should be considered that proof of a statistical correlation between particular indicators does not produce greater clarity.

---

[7]Perceived semiotically, indicators are indexical signs. Cf. p. 99.

It is possible, for example, to prove a positive correlation between an indicator for economic inequality and an indicator for the loss of biodiversity (Mikkelson et al. 2007). Inequality is here measured by the "Gini coefficient of inequality in the distribution of household income"; biodiversity loss is measured through "the number of plant and animal species known to be threatened in 2004". Even if in multiple regression analyses the Gini index of inequality in the distribution of household income can be shown to be a strong predictor variable for biodiversity losses, this still does not reveal an interdependence of the two indicators. A model of the correlation between economic inequality and biodiversity loss would have to demonstrate how specific actions of specific actors can influence, due to an extreme inequality of household income, the number of threatened plant and animal species. Thus, correlations between indicators may provide indications on interdependencies, but they cannot *explain* them.

In global as well as national sustainability politics, indicators are used to provide information to what extent goals have been reached or not. This is also the logic applied by the 2030 Agenda: 17 SDGs with their 169 targets are to be controlled by a total of 304 indicators. Here too, it is important to keep in mind the risks that come with such a sustainability policy.

A rather conspicuous risk is that of questioning a goal retroactively, especially when the indicators already foretell that this goal will not be reached. Less obvious is the risk of self-deception, which is typically comes up when indicators are used not only to gather information about social processes but also to evaluate them politically. Then usually mechanisms are set in motion that aim at producing positive indicator values, largely independently of whether the social processes underlying the indicator values have also taken a positive turn.

This phenomenon was first described by Donald T. Campbell (1979) in the context of educational research. He pointed out that performance tests will only provide objective information about the general performance level of a school, if the teaching and learning are geared toward general competence goals. If instead, teaching and learning are aimed at achieving good results in the performance tests, the results of these tests are useless as indicators.[8] Similar forms of process corruption can be observed in other spheres of social life that are monitored by indicators, and we should be aware of this phenomenon in global environmental and sustainability policy as well. Otherwise, we are in danger of betraying ourselves, since we are mistaking the fulfillment of an indicator for actually reaching a goal.

Models, Simulations, Scenarios

Models, generally speaking, are representations of reality that abstract from the complexity of given conditions, usually with a specific purpose in mind (cf. Stachowiak 1973, 130–133).[9] A scientific model represents *interdependencies* that are crucial for the desired epistemic purpose, always in correspondence with the contemporary perception of what constitutes reality. Additionally, mathematical models can verify how the interdependencies evolve under (controlled) altered conditions. It is important to note that a model always makes the complexities of reality more comprehensible by simplifying them, regardless of whether it only states qualitative differences, points to potential interdependencies, or whether it quantifies its parts as it tests their interdependencies.

In sustainability research, mathematical models and simulations are especially appreciated as they allow researchers to make statements about past developments as well as future possibilities of development.[10] The only other way to objectively determine what results a process will most likely produce is a classic experimental study. Therefore, modeling and simulations are viewed as the predominant methods of experimental science in sustainability research.

---

[8]Campbell (1979, 85) phrased the correlation that he discovered in the following way: "The more any quantitative social indicator is used for social decision-making, the more subject it will be to corruption pressures and the more apt it will be to distort and corrupt the social processes it is intended to monitor".

[9]From a semiotic perspective, models are iconic signs. Cf. p. 99.

[10]In sustainability research, the exemplary work for this kind of knowledge production was the study about "The Limits of Growth" commissioned by the Club of Rome in 1972 (Meadows et al. 1972). For the historical context cf. p. 17ff., especially p. 22.

But with the growing complexity of the observed socio-ecological correlations, it becomes increasingly uncertain whether or not external factors will exert significant influence on future events and developments. Moreover, knowledge gaps are to be expected. This uncertainty is controlled by using scenarios. The vast number of simulation possibilities can be reduced to very few by outlining—through exemplary narratives—plausible interpretations of the basic conditions as they change.

This form of knowledge production is faced with a particular challenge: it needs to explain how the different aspects of social development are adequately taken into account. Here—in addition to the questions already raised—the evaluations entering into any social-ecological modeling process are of crucial importance. These are, on the one hand, technical evaluations determining the degree to which a model is adequate for examining a particular interdependency (cf. Spangenberg 2015; as well as Spangenberg, personal communication, in 2017). Can deterministic models, for instance, be useful to understand stochastic and evolutionary processes?

Beyond such basic decisions, the research process requires constantly renewed evaluations of the social-ecological interdependency. How is it conceptualized? How are physical structures represented, how social structures and how the interactions between them? What data can be used? What data is available? What budget is available for generating, if necessary, new data that is more suitable for the simulations than already existing data? Such evaluations have an immense influence on the research results. Hence, the current discussion about the significance of values (cf. Voinov et al. 2014) and path dependencies in modeling processes (cf. Hämäläinen and Lahtinen 2016). At the same time, standards are being developed for a sound scientific modeling practice, in an effort to render decisions and learning processes transparent and plausible (Grimm et al. 2010; Groeneveld et al. 2012; Müller et al. 2013; Bennett et al. 2013). This discussion is far from over, in fact, is it likely to became even broader, and a closure to the debate is not to be expected soon. Obviously, the scenarios' narratives and the decision-making logic of assessment models are expressions of specific, culturally determined values whose validity may well be questioned, after all.

This methodological strategy bears a risk for environmental and sustainability policy, namely that it is almost impossible to comprehend all the evaluations and decisions included in the modeling process. More transparency in the research process would be of help here. But even then, the complex socio-ecological conditions—which need to be understood in order to be able to shape their future development—would be confronted with complex models that can ultimately only provide answers to specific research questions. Hence, it is doubtful whether this strategy is at all useful for policy analysis (cf. Pindyck 2017).

Another way of dealing with models and scenarios in sustainability discourse is to involve political decision-makers and stakeholders into the modeling process, as clients. Participatory and collaborative modeling approaches can be used to improve the transparency of modeling processes, and to increase the chances that conclusions drawn from the modeling results will contribute to sustainable development. But apart from the problem to produce generalized knowledge from such specific approaches (cf. van Eeten et al. 2002; Sandker et al. 2010; Basco-Carrera et al. 2017), empirical studies are still needed, about whether the scenarios, models, and simulations do in fact gain more validity and practical relevance through client participation (or whether it varies, depending on the stakeholder group involved, or whether the models' validity may even decrease).

Social Science Elements of a Prospective Framework for Research on the Coevolution of Nature and Society

We are part of a society that shapes and modifies what it discovers. Knowledge produced from a coevolutionary perspective is meant to increase the chances that these modifications will not limit the livelihood of humanity, including the livelihood of future generations. Perhaps the phrasing "from a coevolutionary perspective" is a bit stilted. It is supposed to emphasize that searching for coevolutions does not necessarily mean that coevolutions will be discovered. But it helps to expand knowledge if researchers expect that they may discover coevolutionary relations in the first place.

Research about potential coevolutions of nature and society needs adequate access to the subject of the social sciences. After the prior observations, this is at least a postulation to be discussed. Critics often object to such postulations by claiming that it is impossible to speak of *the* social sciences as a unified academic discipline.[11] After all, they like to point out, social scientist neither agree upon a general paradigm shared by a majority of them, nor do they agree that "society" is the central subject of the social sciences (cf. above p. 102).

These observations are correct. The subject of the social sciences is ambiguous, for social scientists are ultimately always a part of the subject they study. They interpret their research object as well as the terms and concepts that represent it in differing contexts and on the basis of varied individual experiences. And with their interpretations they participate—whether they intend to or not—in discourses about the good society or about the good, this is, the right kind of life.

This multiperspectivity is a constitutive characteristic of the social sciences. When researching coevolutions of natural and social phenomena, one needs to acknowledge this fact. For natural scientists, the task is to look for social scientists whose research methods are compatible with their own. Conversely, social scientists who are interested in examining the interdependencies between nature and society together with natural scientists, need in most cases be willing to work with quantitative data. For natural scientists cannot communicate with their research objects quite in the way that social scientists communicate with people as bearers of social structures.

Questions of methodology aside, the concepts we use to explain social phenomena are of crucial importance. Not every concept that is applied in the search for coevolutions is compatible with natural science research perspectives. In many cases, they are exclusively geared toward understanding intentional actions or linguistic practice.

---

[11]Cf. above, p. 35.

The concept of society that I outlined above[12] avoids these limitations. If societies are understood as reflexive sign-using communities, the first look is on the function of symbols, that is, systems of conventional signs. These may be understood as moments of social order. You need language skills, you need to know customs and habits, you need to be familiar with interpretative traditions, in other words, you need knowledge to orient yourself in social situations by means of symbols. Social scientists observe habitual patterns in human behavior to find out *how* this works. With regard to such patterns, society is understood as (open, nonlinear) system whose rules and laws can be reconstructed.[13] Using such basic concepts, the subjects of the social sciences can be understood as coevolutionary variables and it is possible to examine them in cooperation with natural science perspectives.

Additionally, it is essential for a social science perspective to position itself within present-day society. For our view of present-day society corresponds with our assessment of what is worth researching and what not, and with our evaluation of empirical findings. Scientists, for example, who see themselves as part of a capitalist society and draw significant inspiration for their social analyses from the works of Karl Marx and Friedrich Engels, will view things at least partially differently from scientists who view their social conditions as "modern" or "postmodern" and align themselves with corresponding discourses and ways of thinking. Any consideration of social phenomena is embedded in such relations. Social science analyses are characterized not by avoiding these relations but by explicating them.

The basic concept of "society", as outlined above, corresponds with an understanding of present-day society as "modern". Such an understanding of society prevails when the changeability of social structures—and hence their transience—is seen as the driving force of social life (cf. Kaufmann 1989, 38; for a more detailed account, cf. Jetzkowitz 2000, 53–74). When social changes are legitimized by referring to knowledge,

---

[12]Cf. above p. 101ff.

[13]This perspective also allows an investigation of how social systems break away from their routines and develop new structures.

modern societies tend to become so-called knowledge societies (cf. e.g. Stehr 1994). Knowledge societies start to emerge with institutions, which themselves embody the perspectivity of perceptions and evaluations. The subjective assessments expressed through the exchange of money (Simmel 2004) is an example of this process, or interests leading to the formation of political parties, which are competing with one another in democratic power structures (cf. Michels 1949, 134–154). At the same time, these subjective assessments are integrated via procedural law into frameworks of universalistic norms. These are the foundations from which new institutions rise, which—in the role of independent third parties—then review the adequacy of social actions through communicative control and the review of arguments.[14] This process takes on particular forms, depending on the different social areas like science, economy, politics, art, and so on. All these forms have in common, though, that communication and deliberation are used to coordinate action whenever an issue cannot be clearly resolved or a behavioral pattern cannot be conclusively evaluated.

From this perspective of social theory, knowledge and science structures are crucial for decision-making processes about which paths of sustainable social development can be explored and which paths will eventually be chosen. In the following, I first discuss to what extent the existing cognitive and social structures of knowledge production need to be transformed, by means of an analysis of the concept of transdisciplinarity which has been a major influence on the efforts to establish a so-called transformative science (cf. Schneidewind et al. 2016) or a "new global change science" (Pahl-Wostl et al. 2013) and on recent global change research programs (cf. Mooney et al. 2013). Should perhaps the science system itself pursue the goal of bringing about a "great transformation" of social structures? Subsequently, I highlight

---

[14]Cf. Lindemann (2006), who by referring in particular to Simmel and Habermas demonstrates the universal validity of the figure of the third party for a sociological conception of reality (cf. additionally Lindemann 2011). The fact that the figure of the third party in its constitution of symbolic signs is temporally subsequent to the interactions of ego and alter, generates society as a reflexive sign-using community, according to the terminology used here (cf. above p. 96ff.).

coevolutionary science as a possible perspective for knowledge production and discuss the expectations of such a perspective.

Transdisciplinarity and Coevolutionary Science

Among various scholars, the concept of "transdisciplinarity" is considered as a silver bullet to transform global societal structures toward sustainability (cf. e.g. ProClim/CASS 1997; Hirsch Hadorn et al. 2008; Wiesmann et al. 2011; Hurni and Wiesmann 2014). Jean Piaget coined the term in 1970 during an international workshop on "Interdisciplinarity—Teaching and Research Problems in Universities" (Cf. Nicolescu 2005, 1 et seq.; cf. also Nicolescu 2007 for this point and the following argument). He used the term to address the unity of the sciences against the background of the fragmentation of knowledge. The astrophysicist and futurist Erich Jantsch and the French mathematician André Lichnerowicz participated in the discussion during the workshop. Both referred to Piaget's new terminology, but focused on very different aspects (cf. also Bernstein 2015). While Jantsch used the term to argue for innovations in the planning for society at large, Lichnerowicz related it to logic and set theory. After the workshop, "transdisciplinarity" was forgotten until—after a 20-year period of latency—the conceptual debate on transdisciplinarity started up again in the early to mid-1990s. A closer look at the conceptual history and its background helps to understand why the debate resurged, and how to assess the concept's value.

In the early 1970s, the concept of an "environmental research" was established as the scientific response to the environmental problems and the destruction of nature caused by modern industrial societies. This process was accompanied by voices critical of the disciplinary organization of the sciences. To be able to deal adequately with complex problems such as the safeguarding of the possibilities of future developments require the involvement of more than one academic discipline. As a consequence, interdisciplinarity became one of the principal requirements of environmental and science policy, to the point where it was incorporated into funding programs.

The results of these efforts were sobering (cf. e.g. Küppers et al. 1978).[15] Furthermore, environmental research was increasingly superseded by sustainability discourse, so that in the early 1990s the time had come for a new normative program to guide scientific research. The concept, that now was called "transdisciplinarity", asked scientists to abandon their small-minded concentration on homemade research problems of their own disciplines, and to focus instead on solving problems of societal development in cooperation with representatives of social life.

In 1992, Jürgen Mittelstraß reintroduced a newly defined version of Erich Jantsch's concept of "transdisciplinarity" into the on-going debates about science policy (cf. Mittelstraß 1992; Jaeger and Scheringer 1998, 10f. For the history of the concept, cf. esp. Stauffacher 2011, 259–263). At the ETH Zurich, the term was combined with the so-called "mode 2" concept of knowledge production that had been developed by Michael Gibbons, Helga Nowotny, and others (Gibbons et al. 1994; Nowotny et al. 2001). According to this concept, knowledge has been increasingly produced in concrete practical contexts and with a problem-solving focus since the 1950s.[16] Eventually, an international conference, co-organized by Roland Scholz in Zurich in March 2000, was devoted to combining "mode 2" with the concept of transdisciplinarity. The conference led to what is today called the "Zurich school of transdisciplinarity"; its conceptualization of the term essentially dominates the international debate to this day (cf. Bernstein 2015).

A review of its history shows that the primary thrust of the concept of transdisciplinarity is criticism; it reminds scientists of the fact that their work is a social function. The concept brings into a focus a certain discontent with existing structures. The organization of science in disciplines is criticized, as well as the restrictions and regulations that come with it.

---

[15]In Germany, funding for environmental research largely went to established scientists and research teams, who, if necessary, rephrased the language with which they pitched their research topics, but hardly changed their actual research objects, not to mention their ways of thinking.

[16]Mode 1 designates the form of knowledge production initiated by scientists and tied to a scientific discipline. Cf. Gibbons et al. (1994).

Let us take a closer look at the effects of the organization in scientific disciplines. With their specific interdependencies of knowledge and power structures, disciplines regulate what can be explored and communicated within the scientific community.[17] Scientists, for instance, who research environmental problems, need to adjust to the research questions and terminology of their discipline, in order to be able to participate in specialized discussions. Various social control mechanisms guarantee that standards of debate and argumentation in a scientific discipline are met. These can be academic panels deciding about research funding, reviewers of articles for publication in pertinent scientific journals, or even appointment committees for professorships. Whatever these groups do not see as a valid contribution to the specific subject area of their discipline, is rejected.

However, the effects of this kind of academic social control are in fact rather ambivalent. On the one hand, it protects standards of debate and argumentation, on the other hand, it prevents the emergence of fundamental irritations in an established scientific discourse. Such irritations may prompt structural changes, result in new knowledge and help remove barriers to innovation. In other words, fending off irritations means fending off change.

But with the academic control mechanisms firmly a place, scientists whose ideas are not heard are faced with the question of what to. Either they can choose disillusioned withdrawal from scientific discourse, or they can protest, using the language of science, and start, for example, a meta-discussion about the structures of scientific research.

The debate about transdisciplinarity is such a meta-discussion. It is an expression of protest against the established disciplinary structure of academia. Proponents of transdisciplinary research have substantiated their protest in a variety of rather different programs.[18] They all

---

[17]The symbiotic relation of knowledge and power is virtually proverbial. For the correlation of the two, cf. Foucault (1975).

[18]Gibbons et al. (1994, 3–44), for example, proceed descriptively and depict an emergent type of science where knowledge is produced in flexible organizational structures and, in the case of corresponding opportunistic interests, with a focus on its applicability. Conversely, Jaeger and Scheringer (1998) make a more normative argument and demand that transdisciplinary

share a common perception that the professional academic apparatus has become alienated from its social environment: The scientific landscape, so their criticism, has become fragmented into such small, highly specialized discourses that communication across disciplines, as well as communication between the sciences and the field of social practice has become blocked (cf. e.g. Jaeger and Scheringer 1998; Bergmann 2003, 65–67). If this blockade is to be overcome and science's problem-solving competence be restored, then research, so the proponents of transdisciplinarity, needs to focus on where the problems actually are, and not on how they are specifically interpreted within scientific disciplines. To summarize: The development of the concept of transdisciplinarity entails a vision of a science that solves the problems of social practice, without going off on disciplinary tangents, and without losing sight of the social relevance of its research.

It is difficult to say whether this vision can provide the necessary impulse for the scientific landscape, to produce more knowledge relevant to the search for sustainable development paths. That their research results are in one form or other used to shape society, immediately suggests itself at least to social scientists. But going by the current state of the transdisciplinarity debate, it is doubtful that this vision of the sciences can be implemented with the establishment of new, transdisciplinary research institutions and study programs. Transdisciplinary research, on the one hand, can "only" ever be scientific research. Its results need to be transparent and criticizable, and to that end, standards for research and argumentation need to be developed.[19] Transdisciplinary research cannot simply claim to be per se closer to social practice than disciplinary research (cf. Büchi 1996, 205).[20]

---

research be promoted *in addition to* disciplinary, multidisciplinary and interdisciplinary research, to be able to work across the disciplines on problems whose causes are external to the sciences. Schneidewind et al. (2016) developed a "transformative science" program geared toward initiating social change.

[19]For the state of the debate about quality management in transdisciplinary research projects cf. Stauffacher (2011, 264).

[20]Stauffacher (2011, 265–267) likewise emphasizes the limitations of scientific research in transdisciplinary projects, as it relates to their partners involved in social practice. It remains to be seen how more recent approaches, such as transition labs and real-world laboratories

The institutionalization of transdisciplinary research, on the other hand, does not mean that existing disciplinary specializations are simply done away with. Rather, they are reinforced when new fields and specializations of knowledge are developed and even new study programs may be introduced.

Asking for the institutional support and promotion of transdisciplinary research thus comes with unintended consequences. The characteristic features of the established, disciplinary sciences will not disappear even if the proponents of transdisciplinarity were successful. The fragmentation into specialized discourses will even increase. Furthermore, present social structures whose sustainability is doubtful, might be supported by transdisciplinary research and thus survive longer than necessary.[21] Transdisciplinary research, therefore, does not automatically lead to fundamental change and bring about a break with present conditions.

Can these unintended consequences of transdisciplinary research be avoided? For an answer, I think it is useful to look at the methodological core of the concept of transdisciplinarity. Basarab Nicolescu (2005) in particular has called our attention to this. The concept of transdisciplinarity, so Nicolescu, is not to be defined by the function that science provides, or ought to provide, for society. Rather, he develops an interpretation of the concept that is internal to science, an interpretation that refers to general knowledge structures that are valid across all disciplinary specializations.

(cf. Schäpke et al. 2017, 2018), deal with the potential conflict between, on the one hand, support for the idea of social change, and on the other hand, the disinterested examination of change. Depending on the conceptual embeddedness (Schneidewind et al. 2018), different priorities are to be expected. However, tendencies that can be observed in a few projects, namely that they no longer clearly differentiate between the role of agent for a transformative idea and the role of researcher (see, e.g. Pregernig et al. 2018), are reminiscent of the structural problems of action research (see Moser 1978).

[21]Michael Stauffacher (2011, 270; translation J. J.) writes: "Transdisciplinary projects can indeed be understood as neocorporatist negotiation processes, even if significantly expanded by actors from civil society as well as the academic field".

All scientists, regardless of their disciplinary affiliations, says Nicolescu, recognize three methodological principles: The first principle states that there is not only one reality but several layers of reality. The second principle is that of the "logic of the *in*cluded middle", which critically departs from the law of *tertium non datur* of Aristotelian logic. This law states that, for any given proposition either that proposition is true or its negation is true. A third proposition is categorically excluded in the traditions of logic that go back to Aristotle.[22] Not so in Nicolescu's concept of transdisciplinarity: The third principle of transdisciplinarity points to the complexity of the world, in order to avoid that nonlinear and unpredictable, chaotic systems are forced into the straightjacket of a one-dimensional, linear worldview.

Defined like this, transdisciplinarity becomes a complementary concept of disciplinary research.[23] The latter usually deals only with one level of reality. To be sure, scientists are able to experience, in their analyses, that their disciplinary subject is always in complex relations with other aspects of the world. But the methodological norms Nicolescu proposes force scientists to make such experiences, and they thereby establish the unity of the sciences.

Nicolescu's concept is not reductionist, unlike the efforts to create a "unified science" undertaken by the representatives of logical empiricism. Nicolescu neither chooses one science as the foundation of all knowledge, nor does he determine a principle of reality from which all knowledge can be derived.[24] Conversely, he is unable to explain what factually constitutes the unity of science beyond an

---

[22]Incidentally, Aristotle claimed that the logical conclusion inherent in the law of the excluded third (Metaphysics 4.4) is not valid for future events (Peri Hermeneias, Chapter 9).

[23]Cf. DeFreitas et al. (1994), Art. 3: "Transdisciplinarity complements disciplinary approaches".

[24]So far Nicolescu has not submitted a reconstruction of his transdisciplinary principles that would be based on a unified language. Apparently, however, he is not averse to the idea of expressing the transdisciplinary principles in the language of mathematics. Lichnerowicz (cf. Nicolescu 2005, 2) envisioned a similar project. Such a representation—emulating the program of logical empiricism—would be able to claim a foundation for the unity of science that dispenses with any references to metaphysics or a particular *weltanschauung*.

axiomatic methodology. The methodology as well as the 1994 *Charter of Transdisciplinarity* (cf. DeFreitas et al. 1994) formulates specific demands with respect to the ethics of science. But Nicolescu never explains what kind of experiences scientists would have to make to embrace these demands as their own, although he highlights the practice of transdisciplinarity as a test criteria for his axiomatic methodology.[25]

Coevolutionary science has a similar goal as Nicolescu's concept of transdisciplinarity: to encourage scientists to see the bigger picture and to confront their own disciplinary perspectives with knowledge about other subjects, in order to produce new ideas about the world and its interrelations. But it is unlikely that this goal will bring about the proclamation of a new ethics of science. Rather, this goal has to be based on a conceptual research framework which can be argumentatively founded. It must be left to scientific discourse whether such a framework can be integrated with actor–network theory, or—as I have proposed—by searching for explanations of phenomena in research areas where the natural and social sciences intersect, or with the concept of "human interests" (cf. Klintman 2017) or even still other ideas or concepts. Inasmuch as a conceptual framework for coevolutionary research gives equal consideration to the social science perspective and to natural science approaches, it can also incorporate the goal of the Zurich school of transdisciplinarity: namely, to take sustainability discourse and its references to social practice as an incentive to create knowledge, based on the scientific criteria, about the interdependencies between nature and society. Here, a commitment to concrete political goals would negate the diversity of perspectives and limit the possibilities of thinking.[26]

---

[25]"The most important achievement of transdisciplinarity in present times is, of course, the formulation of the methodology of transdisciplinarity, accepted and applied by an important number of researchers in many countries of the world. Transdisciplinarity, in the absence of a methodology, would be just talking, an empty discourse and therefore a short-term living fashion" (Nicolescu 2005, 5).

[26]Cf. in this context, Strohschneider's (2014, 186–190) critical discussion of the concept of transformative science.

Coevolutionary Science as a Perspective for Knowledge Production

Scientists are always a part of society. They do not stand outside of society but in relation to the circumstances in which they live and work. To what degree the individual circumstances of their lives enter into their scientific work, is certainly dependent on their research object and the style of their academic work. But in any case, their view of reality is defined by a certain point of view. For beside their individual circumstances, what scientists see and how they evaluate what they see is also always shaped by their profession.

On what basis, then, can scientists claim to develop knowledge that provides a general orientation about the interdependencies of nature and society? This question leads us straight into the center of the debate about the complex interrelationship between science and society. Without going into the details of this debate, two strategical options to answer this question can be distinguished. Both options rely on rules of behavior, yet aim at the regulation of different aspects of scientific research.

One option is to demand that scientists reflect on how their research is always already shaped by sociocultural contexts, and to refrain from presenting scientific statements as absolutes. To adopt such a position means to emphasize the wide diversity of opinions and claims about the world. Whether or not these opinions and claims can be scientifically justified does not alter the fact that they are, in the end, subjective views. Which opinion becomes the predominant view, is ultimately a question of power. At best, the battle between the bearers of definitional power with their opponents results in social reflexivity (cf. e.g. Beck 1988).

Another strategic option is to demand that scientists explicate the conceptual framework on which they rely to generate and order their perceptions of the (social as well as physical) world. For when scientists document their subjective perceptions of the world within an explicated conceptual framework, they objectify their findings and evaluations.

Coevolutionary science is based on this second strategy. But in light of the above remarks on the debate about transdisciplinarity, it would be naïve not to similarly ask under what conditions coevolutionary science may become popularized and established. Even when the development of a nuanced conceptual framework for the study of interactions between nature and society is indispensable, such a framework will not gain acceptance based on its intellectual potential alone. Science is, after all, not just an epistemological enterprise but also a social system structured by established interests and power structures.

Hence, it is important to promote coevolutionary research, so that studies about the coevolution of nature and society will be conducted and, of course, funded. And, it is also important to demonstrate that knowledge produced from a coevolutionary perspective advances the search for sustainable development paths. In this context, the Intergovernmental Science-Policy Platform on Biodiversity and Ecosystem Services (IPBES) has provided important impulses; in its *Conceptual framework* it has illustrated how nature–society relations can be reimagined (cf. Diaz et al. 2015a, b). Yet, more studies will be necessary, demonstrating again and again that knowledge about the interdependencies of nature and society can indeed be produced in interdisciplinary research. Only when a methodological school has become established, for the collection and classification of knowledge about adequate methods of coevolutionary research, will this be no longer necessary.

Scientists should not be troubled by the fact that a coevolutionary perspective, of the world as a whole and of the interdependencies between nature and society, is "only" one perspective. The times have passed when *the* one science could stylize itself as an objective, outside view of nature and society. Science develops and verifies arguments, which are invariably part of a specific context. When they examine and verify (or reject) reasonable conjectures, scientists explore the structures of the world we live in and bring them into an agreement with the purposes of our actions.[27] And yet, we can never be sure that the path of

---

[27]I emphasized this point in my critical discussion of Luhmann's sceptical attitude. Luhmann's scepticism disregards the indexical dimension of signs. For this, cf. above on pp. 63 and 111.

development chosen by a society will not create existential threats, not for future generations, either. But we can minimize uncertainties when we search for knowledge about potential coevolutionary dynamics and consider these, both in the planning of society and in our personal decisions.

But how can this happen? How does a society have to be organized to accomplish this? And can our present-day societies handle such tasks or do we need completely different social structures?

The polemic phrasing of these questions points in the wrong direction. The last question, in particular, reveals the underlying view, namely that of a radical new beginning: Just imagine that a wholly different society is possible, yes, that such a new society can simply be made to happen—a society with institutions that will not threaten its own sustainability.

The idea of a fundamentally new beginning may at times be an expression of quasi-religious longing, at times a stratagem of political rhetoric. Yet it does not provide a recipe for sustainability discourse, as it contradicts the genetic and tradigenetic modes of the reproduction of life. "There is no *tabula rasa*. We are like sailors who must rebuild their ship at sea, without ever being able to take it apart in a dry-dock and construct it anew from the best parts" (Neurath 1932/1933, 206). This is how Otto Neurath described the human embeddedness in continuous contexts of knowledge in the 1930s. Neurath is often quoted these days. But it is rarely spelled out what this view means with respect to the "difficult-to-grasp idea" (Lee 1993, 13) of a sustainable social development.

Among the proposals for sustainable development paths, one repeatedly finds the demand to reverse the burden of proof when it comes to the introduction of new technologies, in order to hold accountable the producers of the threats to the existence of societies (cf. Ott 1993, 174f.; Beck 1988, 285ff.; Spaemann 1980, 204f.). But so far, there are no answers to the question how such a demand can be reconciled with the values of modern societies. A reversal of the burden of proof encroaches upon the present legal system. In modern society, the legal system is a structure, if not *the* one structure that can give cohesion to a society founded on individual freedom (for an overview of the relevant social theory discourse, cf. Gephardt 1993). Civil liberties are thus an essential element of the law in modern societies. They include the right to research and the right to choose one's profession as well as economic rights. How would a reversal of the burden of proof be compatible with these liberties and the traditions they stand for? Under what conditions could such a change of social habits, in fact, promote sustainable social development? These questions need reliable answers if demands and proposals for sustainable development are to be taken seriously.

Another proposal has been developed by the economic sciences. It relies on the assumption that societies endanger the foundations of their existence if they fail to adequately turn these foundations into objects of barter. From this point of view, the path toward a sustainable future means to internalize external negative costs. Once this is accomplished, everything will be fine (cf. e.g. Endres 2000, 1–34).

Efforts to internalize externalized costs are quite in vogue. Emission trading, for example, is based on this idea. It is assumed that the disposition rights of goods, such as, for example, air, must simply be clearly assigned, in order to internalize the external effects of economic processes.[28] Free access to air is considered an inadequate regime of ownership that should be replaced, by political decision, with a regime of state ownership. By putting caps to the emission of pollutants, air—which from this point on can be used as a carrier of emissions—becomes a scarcer resource. Anyone who wants to or is forced to emit pollutants, may purchase usage rights for a fee and may also further resell these rights—should they no longer be needed. The European Union, for example, established a system for emission trading certificates in 2005 that is supposed to lead to a caring management of ecosystems and a gradual reduction of air pollution (cf. Voß 2007). However, there is no proof yet whether such measures actually result in an effective reduction of pollutant emissions. Other attempts to solve environmental problems with changed disposal rights and the establishment of a market mechanism, urge us to be cautious if not downright skeptical of their possible

---

[28]Most authors here refer to Coase (1960).

success. We have reasons to expect that new markets will produce their own external effects instead of internalizing the external effects of others (cf. Schnaiberg 2005).

Proposals for a (more) sustainable development can only be reliably assessed on the basis of knowledge about potential interrelations with structural decisions. It is here that a new thinking is needed in regards to the production of knowledge. If research about the interdependencies of nature and society can be developed from different conceptions of reality (cf. Barry and Born 2013, 24–26), we need to ask how this diversity can best be accounted for in scientific discourse, but also in public consulting and decision-making procedures. This is important, but as yet neglected area of research, for which the linking of coevolutionary science and, for example, discourse ethics might prove to be groundbreaking (cf. e.g. Jetzkowitz et al. 2017). But in any case, we should say goodbye to idea that there will be one solution or one factor that will redirect the entire development of society toward sustainability. The longing for monocausalities is typical of religions of redemption and their secular offshoots. Scientists may well share this longing, but they should be aware of the fact that it can never be satisfied.

References

Barry, A., & Born, G. (2013). Interdisciplinarity: Reconfigurations of the Social and Natural Sciences. In A. Barry & G. Born (Eds.), *Interdisciplinarity: Reconfigurations of the Social and Natural Sciences* (pp. 1–56). London et al.: Routledge.

Basco-Carrera, L., Warren, A., van Beek, E., Jonoski, A., & Giardino, A. (2017). Collaborative Modelling or Participatory Modelling? A Framework for Water Resources Management. *Environmental Modelling and Software, 91,* 95–110.

Beck, U. (1988). *Gegengifte. Die organisierte Unverantwortlichkeit.* Frankfurt am Main: Suhrkamp.

Bennett, N. D., Croke, B. F. W., Guariso, G., Guillaume, J. H. A., Hamilton, S. H., Jakeman, A. J., et al. (2013). Characterising Performance of Environmental Models. *Environmental Modelling and Software, 40,* 1–20.

Bergmann, M. (2003). Indikatoren für eine diskursive Evaluation transdisziplinärer Forschung. *Technikfolgenabschätzung – Theorie und Praxis, 12*(1), 65–75.

Bernstein, J. (2015). Transdisciplinarity: A Review of Its Origins, Development, and Current Issues. *Journal of Research Practice, 11*(1), Article R1. Retrieved April 21, 2018, from http://jrp.icaap.org/index.php/jrp/article/view/510/412.

Brand, U., & Wissen, M. (2017). *Imperiale Lebensweise. Zur Ausbeutung von Mensch und Natur im globalen Kapitalismus.* München: oekom.

Büchi, H. (1996). Das Paradoxe mit der Transdisziplinarität. *GAIA, 5*(5), 205–208.

Campbell, D. T. (1979). Assessing the Impact of Planned Social Change. *Evaluation and Program Planning, 2*(1), 67–90.

Clark, G. (2007). *A Farewell to Alms: A Brief Economic History of the World.* Princeton: Princeton University Press.

Coase, R. H. (1960). The Problem of Social Cost. *Journal of Law and Economics, 3*(1), 1–44.

Compton, T. J., de Winton, M. D., Leathwick, J. R., & Wadhwa, S. (2012). Predicting Spread of Invasive Macrophytes in New Zealand Lakes Using Indirect Measures of Human Accessibility. *Freshwater Biology, 57,* 938–948.

DeFreitas, L., Morin, E., & Nicolescu, B. (1994). *Charter of Transdisciplinarity. Adopted at the First World Congress of Trandisciplinarity, Convento da Arrábida,* Portugal, November 2–6. Retrieved May 11, 2011, from http://basarab.nicolescu.perso.sfr.fr/ciret/english/charten.htm.

Diaz, S., Demissew, S., Carabias, J., Joly, C., Lonsdale, M., Ash, N., et al. (2015a). The IPBES Conceptual Framework—Connecting Nature and People. *Current Opinion in Environmental Sustainability, 14,* 1–16.

Díaz, S., Demissew, S., Joly, C., Lonsdale, W. M., & Larigauderie, A. (2015b). A Rosetta Stone for Nature's Benefits to People. *PLoS Biology, 13*(1), e1002040. https://doi.org/10.1371/journal.pbio.1002040.

Easterly, W. (2006). *The White Man's Burden: Why the West's Efforts to Aid the Rest Have Done So Much Ill and So Little Good.* New York: Penguin.

Endres, A. (2000). *Umweltökonomie. 3. vollst. überarb u. wes. erw. Aufl.* Stuttgart: Kohlhammer.

Fischer-Kowalski, M., & Weisz, H. (1999). Society as Hybrid Between Material and Symbolic Realms. Towards a Theoretical Framework of Society-Nature-Interaction. *Advances in Human Ecology, 8,* 215–251.

Foucault, M. (1975). *Discipline and Punish: The Birth of the Prison.* New York: Random House.

Fukuda-Parr, S. (2017). *Millenium Development Goals: Ideas, Interests and Influence*. London: Routledge.

Geels, F. W. (2010). Ontologies, Socio-Technical Transitions (to Sustainability), and the Multi-Level Perspective. *Research Policy, 39,* 495–510.

Geels, F. W. (2012a). A Socio-Technical Analysis of Low-Carbon Transitions: Introducing the Multi-Level Perspective into Transport Studies. *Journal of Transport Geography, 24,* 471–482.

Geels, F. W. (2012b). The Multi-Level Perspective on Sustainability Transitions: Responses to Seven Criticisms. *Environmental Innovation and Societal Transitions, 1*(1), 24–40.

Geels, F. W., & Schot, J. (2007). Typology of Sociotechnical Transition Pathways. *Research Policy, 36*(3), 399–417.

Gephardt, W. (1993). *Gesellschaftstheorie und Recht: Das Recht im soziologischen Diskurs der Moderne*. Frankfurt am Main: Suhrkamp.

Gibbons, M., Limoges, C., Nowotny, H., Schwartzmann, S., Scott, P., & Trow, M. (1994). *The New Production of Knowledge. The Dynamics of Science and Research in Contemporary Societies*. London: Sage.

Grimm, V., Berger, U., DeAngelis, D. L., Polhill, J. G., Giske, J., & Railsback, S. F. (2010). The ODD Protocol: A Review and First Update. *Ecological Modelling, 221,* 2760–2768.

Groeneveld, J., Müller, B., Angermüller, F., Drees, R., Dreßler, G., Klassert, C., et al. (2012). Good Modelling Practice: Expanding the ODD Model Description Protocol for Socioenvironmental Agent Based Models. In R. Seppelt, A. A. Voinov, S. Lange, & D. Bankamp (Eds.), *Managing Resources of a Limited Planet: Pathways and Visions Under Uncertainty. Proceedings of the Sixth Biennial Meeting of the International Environmental Modelling and Software Society*, Leipzig, Germany, July 1–5, 2012 (pp. 2480–2484). Manno: iEMSs.

Hämäläinen, R. P., & Lahtinen, T. J. (2016). Path Dependence in Operational Research—How the Modeling Process Can Influence the Results. *Operations Research Perspectives, 3,* 14–20.

Haraway, D. (1985). A Manifesto for Cyborgs: Science, Technology, and Socialist Feminism in the 1980s. *Socialist Review, 80,* 65–108.

Hirsch Hadorn, G., Hoffmann-Riem, H., Biber-Klemm, S., Grossenbacher-Mansuy, W., Joye, D., Pohl, C., et al. (Eds.). (2008). *Handbook of Transdisciplinary Research*. Berlin: Springer.

Hurni, H., & Wiesmann, U. M. (2014). Transdisciplinarity in Practice. Experience from a Concept-Based Research Programme Addressing Global Change and Sustainable Development. *GAIA, 23*(3), 275–277.

Jaeger, J., & Scheringer, M. (1998). Transdisziplinarität: Problemorientierung ohne Methodenzwang. *GAIA, 7*(1), 10–25.

Jetzkowitz, J. (2000). *Recht und Religion in der modernen Gesellschaft: Soziologische Theorie und Analyse am Beispiel der Rechtsprechung des Bundesverfassungsgerichts in Sachen "Religion" zwischen den Jahren 1983 und 1997.* Münster: LIT-Verlag.

Jetzkowitz, J., van Koppen, C. S. A. (Kris), Lidskog, R., Ott, K., Voget-Kleschin, L., & Mei Ling Wong, C. (2017). The Significance of Meaning: Why IPBES Needs the Social Sciences and Humanities. *Innovation: The European Journal of Social Science Research, 31*(suppl.1), S38–S60. https://doi.org/10.1080/13511610.2017.1348933.

Kaufmann, F.-X. (1989). *Religion und Modernität: Sozialwissenschaftliche Perspektiven.* Tübingen: J. C. B. Mohr (Paul Siebeck).

Kizuka, T., Akasaka, M., Kadoya, T., & Takamura, N. (2014). Visibility from Roads Predict the Distribution of Invasive Fishes in Agricultural Ponds. *PLoS One, 9*(6), e99709. https://doi.org/10.1371/journal.pone.0099709.

Klintman, M. (2017). *Human Sciences and Human Interests: Integrating the Social, Economic, and Evolutionary Sciences.* London and New York: Routledge.

Küppers, G., Lundgreen, P., & Weingart, P. (1978). *Umweltforschung – die gesteuerte Wissenschaft? Eine empirische Studie zum Verhältnis von Wissenschaftsentwicklung und Wissenschaftspolitik.* Frankfurt am Main: Suhrkamp.

Latour, B. (2004). *Politics of Nature: How to Bring the Sciences into Democracy* (P. Catherine, Trans.). Cambridge, MA: Harvard University Press.

Le Blanc, D. (2015). *Towards Integration at Last? The Sustainable Development Goals as a Network of Targets, United Nations* (DESA Working Paper No. 141). Retrieved April 16, 2018, from https://www.un.org/esa/desa/papers/2015/wp141_2015.pdf.

Lee, K. N. (1993). *Compass and Gyroscope: Integrating Science and Politics for the Environment.* Washington, DC: Island Press.

Lessenich, S. (2017). *Neben uns die Sintflut: Die Externalisierungsgesellschaft und ihr Preis.* Berlin: Hanser.

Lindemann, G. (2006). Die Emergenzfunktion und die konstitutive Funktion des Dritten. Perspektiven einer kritisch-systematischen Theorieentwicklung. *Zeitschrift für Soziologie, 35*(2), 82–101.

Lindemann, G. (2011). Die Gesellschaftstheorie von der Sozialtheorie her denken – oder umgekehrt? *ZfS-Forum, 3*(1), 1–19. Retrieved March 31, 2014, from http://www.zfs-online.org/index.php/forum/article/viewFile/3066/2603.

Meadows, D. L., Meadows, D. H., Randers, J., & Behrens III, W. W. (1972). *The Limits to Growth: A Report for the Club of Rome's Project on the Predicament of Mankind.* New York: Universe Books.

Michels, R. (1949). *Political Parties.* Glencoe, IL: Free Press.

Mikkelson, G. M., Gonzales, A., & Peterson, G. D. (2007). Economic Inequality Predicts Biodiversity Loss. *PLoS One, 2*(5), e444. https://doi.org/10.1371/journal.pone.0000444.

Mittelstraß, J. (1992). Auf dem Wege zur Transdisziplinarität. *GAIA, 1*(5), 250.

Mol, A. P. J., & Spaargaren, G. (2006). Towards a Sociology of Environmental Flows: A New Agenda for 21st Century Environmental Sociology. In G. Spargaaren, A. P. J. Mol, & F. H. Buttel (Eds.), *Governing Environmental Flows: Global Challenges to Social Theory* (pp. 39–82). Cambridge, MA: MIT Press.

Mooney, H. A., Duraiappah, A., & Larigauderie, A. (2013). Evolution of Natural and Social Science Interactions in Global Change Research Programs. *Proceedings of the National Academy of Sciences,* January 2013 (201107484). https://doi.org/10.1073/pnas.1107484110.

Moore, J. W. (2015). *Capitalism in the Web of Life: Ecology and the Accumulation of Capital.* London et al.: Verso.

Moser, H. (1978). Einige Aspekte der Aktionsforschung im internationalen Vergleich. In H. Moser & H. Ornauer (Eds.), *Internationale Aspekte der Aktionsforschung* (pp. 173–189). München: Kösel.

Müller, B., Bohn, F., Dreßler, G., Groeneveld, J., Klassert, C., Martin, R., et al. (2013). Describing Human Decisions in Agent-Based Models—ODD + D, an Extension of the ODD Protocol. *Environmental Modelling and Software, 48,* 37–48.

Neurath, O. (1932/1933). Protokollsätze. *Erkenntnis, 3*(1), 204–214 [English Translation in: R. S. Cohen & M. Neurath (Eds.), *Philosophical Papers 1913–1946. Vienna Circle Collection* (Vol. 16, pp. 91–99). Dordrecht: Springer].

Nicolescu, B. (2005). *Transdisciplinarity—Past, Present and Future.* Retrieved March 28, 2011, from http://cetrans.com.br/textos/transdisciplinarity-past-present-and-future.pdf.

Nicolescu, B. (2007). *Transdisciplinarity as Methodological Framework for Going Beyond the Science-Religion Debate.* Retrieved March 28, 2011, from http://www.fernandosantiago.com.br/transdisx.pdf.

Nilsson, M., Griggs, D., & Visbeck, M. (2016). Map the Interactions Between Sustainable Development Goals. *Nature, 534,* 320–322.

Norgaard, R. B. (1994). *Development Betrayed: The End of Progress and a Coevolutionary Revisioning of the Future*. London: Routledge.

Nowotny, H. (2001). Vom Geschichtenerzählen zur Koevolutionswissenschaft. *GAIA, 10*(4), 262–264.

Ostrom, E. (2009). A General Framework for Analyzing Sustainability of Social-Ecological Systems. *Science, 325*, 419–422.

Ott, K. (1993). *Ökologie und Ethik: Ein Versuch praktischer Philosophie*. Tübingen: Attempto-Verlag.

Pahl-Wostl, C., Giupponi, C., Richards, K., Binder, C., de Sherbinin, A., Sprinz, D., et al. (2013). Transition Towards a New Global Change Science: Requirements for Methodologies, Methods, Data and Knowledge. *Environmental Science & Policy, 28*, 36–47.

Pindyck, R. S. (2017). The Use and Misuse of Models for Climate Policy. *Review of Environmental Economics and Policy, 11*(1), 100–114.

Pradhan, P., Costa, L., Rybski, D., Lucht, W., & Kropp, J. P. (2017). A Systematic Study of Sustainable Development Goal (SDG) Interactions. *Earth's Future, 5*, 1169–1179.

Pregernig, M., Rhodius, R., & Winkel, G. (2018). Design Junctions in Real-World Laboratories: Analyzing Experiences Gained from the Project Knowledge Dialogue Northern Black Forest. *GAIA, 27*(S1), 32–38.

ProClim/CASS, Konferenz der Schweizerischen Wissenschaftlichen Akademien. (Eds.). (1997). *Forschung zu Nachhaltigkeit und globalem Wandel – wissenschaftspolitische Visionen der schweizer Forschenden*. Bern: ProClim.

Sandker, M., Campbell, B. M., Ruiz-Pérez, M., Sayer, J. A., Cowling, R., Kassa, H., & Knight, A. T. (2010). The Role of Participatory Modeling in Landscape Approaches to Reconcile Conservation and Development. *Ecology and Society, 15*(2), 13. Retrieved April 20, 2018, from http://www.ecologyandsociety.org/vol15/iss2/art13/.

Schäpke, N., Stelzer, F., Caniglia, G., Bergmann, M., Wanner, M., Singer-Brodowski, M., et al. (2018). Jointly Experimenting for Transformation? Shaping Real-World Laboratories by Comparing Them. *GAIA, 27*(S1), 85–96.

Schäpke, N., Stelzer, F., Marg, O., Bergmann, M., Miller, E., Wagner, F., et al. (2017). Urban BaWü-Labs: Challenges and Solutions When Expanding the Real-World Lab Infrastructure. *GAIA, 26*(4), 366–368.

Schnaiberg, A. (2005). The Economy and the Environment. In N. J. Smelser & R. Swedberg (Eds.), *The Handbook of Economic Sociology* (2nd ed., pp. 703–725). Princeton: Princeton University Press.

Schneidewind, U., Augenstein, K., Stelzer, F., & Wanner, M. (2018). Structure Matters: Real-World Laboratories as a New Type of Large-Scale Research Infrastructure—A Framework Inspired by Giddens' Structuration Theory. *GAIA, 27*(S1), 12–17.

Schneidewind, U., Singer-Brodowski, M., Augenstein, K. (2016). Transformative Science for Sustainability Transitions. In H. G. Brauch, U. Oswald Spring, J. Grin, & J. Scheffran (Eds.), *Handbook on Sustainability Transition and Sustainable Peace* (pp. 123–136). Berlin and Heidelberg: Springer.

Shove, E. (2010). Beyond the ABC: Climate Change Policy and Theories of Social Change. *Environment and Planning A, 42*(6), 1273–1285.

Simmel, G. (2004 [1900]). *The Philosophy of Money* (Edieted by David Frisby. Translated by Tom Bottomore and David Frisby from a First Draft by Kaethe Mengelberg). London and New York: Routledge.

Spaemann, R. (1980). Technische Eingriffe in die Natur als Problem der politischen Ethik. In D. Birnbacher (Ed.), *Ökologie und Ethik* (pp. 180–206). Stuttgart: Reclam.

Spangenberg, J. (2015). Sustainability and the Challenge of Complex Systems. In J. C. Enders, & M. Remig (Eds.), *Theories of Sustainable Development* (pp. 89–111). London and New York: Routledge.

Stachowiak, H. (1973). *Allgemeine Modelltheorie*. Wien and New York: Springer.

Stauffacher, M. (2011). Umweltsoziologie und Transdisziplinarität. In M. Groß (Ed.), *Handbuch Umweltsoziologie* (pp. 259–276). Wiesabaden: VS Verlag.

Stehr, N. (1994). *Knowledge Societies*. London: Sage.

Strohschneider, P. (2014). Zur Politik der Transformativen Wissenschaft. In A. Brodocz, D. Herrmann, R. Schmidt, D. Schulz, & J. Schulze-Wessel (Eds.), *Die Verfassung des Politischen. Festschrift für Hans Vorländer* (pp. 171–192). Wiesbaden: Springer.

van Eeten, M. J. G., Loucks, D. P., & Roe, E. (2002). Bringing Actors Together Around Large-Scale Water Systems: Participatory Modeling and Other Innovations. *Knowledge, Technology, & Policy, 14*(4), 94–108.

Voß, J.-P., Newig, J., Kastens, B., Monstadt, J., & Nöltig, B. (2007). Steering for Sustainable Development: A Typology of Problems and Strategies with Respect to Ambivalence, Uncertainty and Distributive Power. *Journal of Environmental Policy & Planning, 9*(3/4), 193–212.

Voinov, A., Seppelt, R., Reis, S., Nabel, J. E. M. S., & Shokravi, S. (2014). Values in Socio-Environmental Modelling: Persuasion for Action or Excuse for Inaction. *Environmental Modelling and Software, 53,* 207–212.

Wallerstein, I. M. (1974–1989). *The Modern World-System* (3 vols.). New York, London and San Diego, CA: Academic Press.

Wiesmann, U., Hurni, H., Ott, C., & Zingerli, C. (2011). Combining the Concepts of Transdisciplinarity and Partnership in Research for Sustainable Development. In U. Wiesmann & H. Hurni (Eds.), *Research for Sustainable Development: Foundations, Experiences, and Perspectives. Perspectives of the Swiss National Centre of Competence in Research (NCCR) North-South* (pp. 43–70). Bern: Geographica Bernensia.

Williams, E. E. (1944). *Capitalism and Slavery.* Chapel Hill: University of North Carolina Press.

References

Agamben, G. (1998 [1995]). *Homo Sacer: Sovereign Power and Bare Life* (D. Heller-Roazen, Trans.). Stanford: Stanford University Press.

Alexander, J. C. (1987). Action and Its Environments. In J. C. Alexander, B. Giesen, R. Münch, & N. J. Smelser (Eds.), *The Micro-Macro Link* (pp. 289–318). Berkeley et al: University of California Press.

Aristoteles. (1978). Politik – griechisch und deutsch, ed. F. Susemihl, Part I: Text und Übersetzung. In Aristoteles, Werke, Vol. 6. Reprint of the edition Leipzig 1879, Aalen: Scientia-Verlag.

Aristoteles. (1989). Metaphysik – griechisch-deutsch; revision of the translation by Hermann Bonitz, edited with an introduction and comments by Horst Seidl; Greek text by Wilhelm Christ. 3rd revised edition, Hamburg: Meiner.

Aristoteles. (2002). *Peri Hermeneias* (H. Weidemann, Trans. and Explained). Berlin: Akademie-Verlag.

Avital, E., & Jablonka, E. (2000). *Animal Traditions: Behavioural Inheritance in Evolution*. Cambridge: Cambridge University Press.

Barry, A., & Born, G. (2013). Interdisciplinarity: Reconfigurations of the Social and Natural Sciences. In A. Barry & G. Born (Eds.), *Interdisciplinarity: Reconfigurations of the Social and Natural Sciences* (pp. 1–56). London et al.: Routledge.

Basco-Carrera, L., Warren, A., van Beek, E., Jonoski, A., & Giardino, A. (2017). Collaborative Modelling or Participatory Modelling? A Framework for Water Resources Management. *Environmental Modelling & Software, 91,* 95–110.

Bateson, G. (2000 [1972]). Steps to an Ecology of Mind: Collected Essays in Anthropology, Psychiatry, Evolution, and Epistemology. Chicago, IL: University of Chicago Press.

Beck, U. (1988). *Gegengifte Die organisierte Unverantwortlichkeit.* Frankfurt am Main: Suhrkamp.

Beck, U. (1992 [1986]). Risk Society: Towards a New Modernity (M. Ritter, Trans.). London: Sage [originally published in German: Beck, U. (1986). *Risikogesellschaft. Auf dem Weg in eine andere Moderne.* Frankfurt am Main: Suhrkamp].

Beck, U. (1993). *Die Erfindung des Politischen.* Frankfurt am Main: Suhrkamp.

Beck, U. (1999). *World Risk Society.* Cambridge: Polity Press.

Beck, U., Giddens, A., & Lash, S. (1996). *Reflexive Modernisierung Eine Kontroverse.* Suhrkamp: Frankfurt am Main.

Beckert, J. (1997). *Grenzen des Marktes. Die sozialen Grundlagen wirtschaftlicher Effizienz.* Frankfurt am Main and New York: Campus.

Bennett, N. D., Croke, B. F. W., Guariso, G., Guillaume, J. H. A., Hamilton, S. H., Jakeman, A. J., et al. (2013). Characterising Performance of Environmental Models. *Environmental Modelling and Software, 40,* 1–20.

Bergmann, M. (2003). Indikatoren für eine diskursive Evaluation transdisziplinärer Forschung. *Technikfolgenabschätzung – Theorie und Praxis, 12*(1), 65–75.

Berkes, F., Colding, J., & Folke, C. (Eds.). (2003). *Navigating Social-Ecological Systems. Building Resilience for Complexity and Change.* Cambridge: Cambridge University Press.

Bernstein, J. (2015). Transdisciplinarity: A Review of Its Origins, Development, and Current Issues. *Journal of Research Practice, 11*(1), Article R1. Retrieved April 21, 2018, from http://jrp.icaap.org/index.php/jrp/article/view/510/412.

Bettencourt, L. M. A., & Kaur, J. (2011). Evolution and Structure of Sustainability Science. *PNAS, 108*(49), 19540–19545.

Böhme, G. (1997). Natur. In C. Wulff (Hg.), *Vom Menschen. Handbuch Historische Anthropologie* (pp. 92–116). Weinheim und Basel: Beltz.

Böhme, G., van den Daele, W., & Krohn, W. (1972). Alternativen in der Wissenschaft. *Zeitschrift für Soziologie, 1*(4), 302–316.

Böhme, G., van den Daele, W., & Krohn, W. (1973). Die Finalisierung der Wissenschaft. *Zeitschrift für Soziologie, 2*(2), 128–144.

Böhret, C., & Konzendorf, G. (1997). *Ko-Evolution von Gesellschaft und funktionalem Staat. Ein Beitrag zur Theorie der Politik.* Opladen: Westdeutscher Verlag.

Bonner, J. T. (1980). The Evolution of Culture in Animals. Princeton, NJ: Princeton University Press.

Boole, G. (1854). *An Investigation of the Laws of Thought, on Which Are Founded the Mathematical Theories of Logic and Probabilities.* New York: Dover.

Bourdieu, P. (Ed.). (1997). *The Weight of the World: Social Suffering in Contemporary Society.* Cambridge: Polity.

Brand, U., & Wissen, M. (2017). *Imperiale Lebensweise. Zur Ausbeutung von Mensch und Natur im globalen Kapitalismus.* München: oekom.

Brandom, R. B. (1994). *Making It Explicit: Reasoning, Representing, and Discursive Commitment.* Cambridge, MA: Harvard University Press.

Brandom, R. B. (2001). *Begründen und Begreifen. Eine Einführung in den Inferentialismus.* Suhrkamp: Frankfurt am Main.

Bricmont, J., & Sokal, A. (1999). *Fashionable Nonsense: Postmodern Intellectuals' Abuse of Science.* New York: Picador.

Brulle, R. J. (2000). *Agency, Democracy, and Nature: The U.S. Environmental Movement Organizations from a Critical Theory Perspective.* Cambridge, MA: MIT Press.

Bryson, B. (2003). *A Short History of Nearly Everything.* New York: Random House.

Büchi, H. (1996). Das Paradoxe mit der Transdisziplinarität. *GAIA, 5*(5), 205–208.

Burger, P. (2006). Sustainability Science: The Science of the Future. Unpublished Manuscript.

Burger, P. (2007). Nachhaltigkeitstheorie als Gesellschaftstheorie. Ein philosophisches Plädoyer. In Schweizerische Akademie der Geistes- und Sozialwissenschaften (Ed.), *Nachhaltigkeitsforschung – Perspektiven der Sozial- und Geisteswissenschaften* (pp. 13–34). Bern: Schweizerische Akademie der Geistes- und Sozialwissenschaften.

Burger, P., & Christen, M. (2011). Towards a Capability Approach of Sustainability. *Journal of Cleaner Production, 19*(8), 787–795.

Burkholz, R. (2008). *Problemlösende Argumentationsketten.* Weilerwist: Velbrück.

Burningham, K. (1998). A Noisy Road or Noisy Resident?: A Demonstration of the Utility of Social Constructionism for Analysing Environmental Problems. *The Sociological Review, 46*(3), 536–563.

Burningham, K., & Cooper, G. (1999). Being Constructive: Social Constructionism and the Environment. *Sociology, 33*(2), 297–316.

Campbell, D. T. (1979). Assessing the Impact of Planned Social Change. *Evaluation and Program Planning, 2*(1), 67–90.

Carson, R. (1962). *Silent Spring*. Greenwich, CT: Fawcett.

Caspar, M. (1937). Nachbericht. In J. Kepler, 1937: M. Caspar (Ed.), *Gesammelte Werke: Vol. 3. Astronomia Nova* (pp. 427–484). München: Beck.

Caspar, M. (1995). *Johannes Kepler*. Stuttgart: Kohlhammer.

Clark, G. (2007). *A Farewell to Alms: A Brief Economic History of the World*. Princeton: Princeton University Press.

Coase, R. H. (1960). The Problem of Social Cost. *Journal of Law and Economics, 3*(1), 1–44.

Commoner, B. (1971). *The Closing Circle: Nature, Man, and Technology*. New York: Knopf.

Compton, T. J., de Winton, M. D., Leathwick, J. R., & Wadhwa, S. (2012). Predicting Spread of Invasive Macrophytes in New Zealand Lakes Using Indirect Measures of Human Accessibility. *Freshwater Biology, 57*, 938–948.

Comte, A. (1903 [1844]). *A Discourse on the Positive Spirit* (E. S. Beesly, Trans.). London: William Reeves.

Connor, J. A. (2004). *Kepler's Witch. An Astronomer's Discovery of Cosmic Order Amid Religious War, Political Intrigue, and the Heresy Trial Of His Mother*. San Francisco, CA: Harper.

Cortekar, J., Jasper, J., & Sundmacher, T. (2006). *Die Umwelt in der Geschichte des ökonomischen Denkens*. Marburg: Metropolis-Verlag.

Costanza, R., Cumberland, J., Daly, H., Goodland, R., & Norgaard, R. (1997). *An Introduction to Ecological Economics*. Boca Raton, FL: CRC Press LLC.

Dahrendorf, R. (1989). Einführung in die Soziologie. *Soziale Welt, 40*(1/2), 2–10.

Daly, H. E. (1973). The Steady State Economy: Toward a Political Economy of Biophysical Equilibrium and Moral Growth. In H. E. Daly (Ed.), *Toward a Steady State Economy* (pp. 149–174). San Francisco: W.H. Freeman.

Daly, H. E. (1987). The Economic Growth Debate: What Some Economists Have Learned but Many Have Not. *Journal of Environmental Economics and Management, 14*, 323–336.

Daly, H. E. (2008). *A Steady-State Economy.* Report to the Sustainable Development Commission, UK (April 24, 2008). Retrieved November 16, 2013, from http://www.sd-commission.org.uk/data/files/publications/Herman_Daly_thinkpiece.pdf.

Danneberg, L. (1988). Peirces Abduktionskonzeption als Entdeckungslogik. Eine philosophiehistorische und rezeptionskritische Untersuchung. *Archiv für Geschichte der Philosophie, 70,* 305–326.

Danneberg, L. (1989). *Methodologien: Struktur, Aufbau und Evaluation.* Berlin: Duncker & Humblot.

DeFreitas, L., Morin, E., & Nicolescu, B. (1994). *Charter of Transdisciplinarity. Adopted at the First World Congress of Trandisciplinarity, Convento da Arrábida,* Portugal, November 2–6. Retrieved May 11, 2011, from http://basarab.nicolescu.perso.sfr.fr/ciret/english/charten.htm.

Dennett, D. C. (1991). *Consciousness Explained.* Boston: Little, Brown and Co.

Diaz, S., Demissew, S., Carabias, J., Joly, C., Lonsdale, M., Ash, N., et al. (2015a). The IPBES Conceptual Framework—Connecting Nature and People. *Current Opinion in Environmental Sustainability, 14,* 1–16.

Diaz, S., Demissew, S., Joly, C., Lonsdale, W. M., & Larigauderie, A. (2015b). A Rosetta Stone for Nature's Benefits to People. *PLoS Biology, 13*(1), e1002040. https://doi.org/10.1371/journal.pbio.1002040.

Dickens, P. (2004). *Society and Nature: Changing Our Environment, Changing Ourselves.* Cambridge: Polity Press.

Didham, R. J., & Ofei-Manu, P. (2015). Social Learning for Sustainability. In V. W. Thoresen, D. Doyle, J. Klein, & R. J. Didham (Eds.), *Responsible Living: Concepts, Education and Future Perspectives* (pp. 233–252). Cham: Springer.

Dryzek, J. S. (1987). *Rational Ecology, Environment and Political Economy.* Oxford: Basil Blackwell.

Durkheim, E. (1995 [1895]). Die Regeln der soziologischen Methode (R. König, Ed. and Intro.). Frankfurt am Main: Suhrkamp.

Durkheim, E. (1996 [1893]). *Über soziale Arbeitsteilung. Studie über die Organisation höherer Gesellschaften.* Frankfurt am Main: Suhrkamp.

Dyball, R., Brown, V. A., & Keen, M. (2007). Towards Sustainability: Five Strands of Social Learning. In A. E. J. Wals (Ed.), *Social Learning: Towards a Sustainable World* (pp. 181–194). Wageningen: Wageningen Academic Publishers.

Easterly, W. (2006). *The White Man's Burden: Why the West's Efforts to Aid the Rest Have Done So Much Ill and So Little Good.* New York: Penguin.

Eckersley, R. (1992). *Environmentalism and Political Theory: Toward an Ecocentric Approach*. Albany, NY: Suny Press.

Eckersley, R. (1994). Wo bleibt die Emanzipation der Natur? Habermas' kritische Theorie aus ökozentrischer Sicht. In W. Zierhofer & D. Steiner (Eds.), *Vernunft angesichts der Umweltzerstörung* (pp. 119–158). Opladen: Westdeutscher Verlag.

Eco, U. (1983). Horns, Hooves, Insteps: Some Hypotheses on Three Types of Abduction. In U. Eco & T. A. Sebeok (Eds.), *The Sign of Three—Dupin, Holmes, Peirce* (pp. 198–220). Bloomington: Indiana University Press.

Eder, K. (1976). *Die Entstehung staatlich organisierter Gesellschaften. Ein Beitrag zu einer Theorie sozialer Evolution.* Suhrkamp: Frankfurt am Main.

Eder, K. (1999). Societies Learn and Yet the World is Hard to Change. *European Journal of Social Theory, 2*, 195–215.

Ehrlich, P. R., & Raven, P. H. (1964). Butterflies and Plants: A Study in Coevolution. *Evolution, 18*, 586–608.

Einstein, A., & Infeld, L. (1956). *Die Evolution der Physik. Von Newton bis zur Quantentheorie.* Reinbek bei Hamburg: Rowohlt.

Elias, N. (1970). *Was ist Soziologie?* München: Juventa [also published in English: Elias, 1978: What Is Sociology? London: Hutchinson].

Elias, N. (1977). Zur Grundlegung einer Theorie sozialer Prozesse. *Zeitschrift für Soziologie, 6*(2), 127–149 [also published in English: Elias, 1997: Towards a Theory of Social Processes. *British Journal of Sociology, 48*(3), 355–383].

Elias, N. (1998 [1939]). *Über den Prozeß der Zivilisation* (2., rev. and expanded ed.). Frankfurt am Main: Suhrkamp [also published in English: Elias. (2012). *On the Process of Civilisation*, ed. S. Mennell, E. Dunning, J. Goudsblom, & R. Kilminster. Dublin: UCD Press].

Emirbayer, M., & Mische, A. (1998). What is Agency? *American Journal of Sociology, 103*(4), 962–1023.

Endres, A. (2000). *Umweltökonomie. 3. vollst. überarb u. wes. erw. Aufl.* Stuttgart: Kohlhammer.

Engelhardt, A., & Kajetzke, L. (Eds.). (2010). *Handbuch Wissensgesellschaft. Theorien, Themen und Probleme.* Bielefeld: transcript.

Engels, F. (1979 [1892]). Engels an Laura Lafargue in Le Perreux. In K. Marx & F. Engels, *Werke* (Band 38, pp. 544–546). Berlin: Dietz-Verlag.

Esfeld, M. (2002). *Einführung in die Naturphilosophie.* Darmstadt: Wissenschaftliche Buchgesellschaft.

Esser, H. (1996). *Soziologie. Allgemeine Grundlagen.* Frankfurt am Main and New York: Campus.

Fischer, H., Kumke, T., Lohmann, G., Flöser, G., Miller, H., von Storch, H., & Negendank, J. F. W. (Eds.). (2004). The Climate in Historical Times. Towards a Synthesis of Holocene Proxy Data and Climate Models. In *Proceedings of the Third GKSS School on Environmental Research*. Berlin: Springer.

Fischer-Kowalski, M., & Weisz, H. (1999). Society as Hybrid Between Material and Symbolic Realms. Towards a Theoretical Framework of Society-Nature-Interaction. *Advances in Human Ecology, 8,* 215–251.

Flannery, T. (2005). *The Weather Makers: The History and Future Impact of Climate Change*. London: Allen Lane.

Foucault, M. (1975). *Discipline and Punish: The Birth of the Prison*. New York: Random House.

Foukal, P., Fröhlich, C., Spruit, H., & Wigley, T. M. L. (2006). Variations in Solar Luminosity and Their Effect on the Earth's Climate. *Nature, 443,* 161–166.

Frankfurter Rundschau, 4 July 2006, 1.

Fukuda-Parr, S. (2017). *Millenium Development Goals: Ideas, Interests and Influence*. London: Routledge.

Gay, P. (1967/1970). The Enlightenment: An Interpretation. London: Weidenfeld & Nicolson.

Geels, F. W. (2010). Ontologies, Socio-Technical Transitions (to Sustainability), and the Multi-Level Perspective. *Research Policy, 39,* 495–510.

Geels, F. W. (2012a). A Socio-Technical Analysis of Low-Carbon Transitions: Introducing the Multi-Level Perspective into Transport Studies. *Journal of Transport Geography, 24,* 471–482.

Geels, F. W. (2012b). The Multi-Level Perspective on Sustainability Transitions: Responses to Seven Criticisms. *Environmental Innovation and Societal Transitions, 1*(1), 24–40.

Geels, F. W., & Schot, J. (2007). Typology of Sociotechnical Transition Pathways. *Research Policy, 36*(3), 399–417.

Gephardt, W. (1993). *Gesellschaftstheorie und Recht: Das Recht im soziologischen Diskurs der Moderne*. Frankfurt am Main: Suhrkamp.

Gibbons, M., Limoges, C., Nowotny, H., Schwartzmann, S., Scott, P., & Trow, M. (1994). *The New Production of Knowledge. The Dynamics of Science and Research in Contemporary Societies*. London: Sage.

Giesen, B. (1999). Codes kollektiver Identität. In W. Gephart & H. Waldenfels (Eds.), *Religion und Identität. Im Horizont des Pluralismus* (pp. 13–43). Frankfurt am Main: Suhrkamp.

Gigerenzer, G. (1991). From Tools to Theories: A Heuristic of Discovery in Cognitive Psychology. *Psychological Review, 98*(2), 254–267.

Gilder, J., & Gilder, A.-L. (2004). *Heavenly Intrigue. Johannes Kepler, Tycho Brahe, and the murder behind one of history's greatest scientific discoveries.* New York: Doubleday.

Glotfelty, C., & Fromm, H. (1996). *The Ecocriticism Reader: Landmarks in Literary Ecology.* Athens and London: University of Georgia Press.

Gloy, K. (1995). *Die Geschichte des wissenschaftlichen Denkens. Verständnis der Natur.* Beck: München.

Görg, C. (1999). *Gesellschaftliche Naturverhältnisse.* Münster: Westfälisches Dampfboot.

Grimm, V., Berger, U., DeAngelis, D. L., Polhill, J. G., Giske, J., & Railsback, S. F. (2010). The ODD Protocol: A Review and First Update. *Ecological Modelling, 221,* 2760–2768.

Grober, U. (2010). *Die Entdeckung der Nachhaltigkeit. Kulturgeschichte eines Begriffs.* München: Verlag Antje Kunstmann. [also published in English: Grober, U. (2012). Sustainability: A Cultural History (R. Cunningham, Trans.). Totnes, Devon: Green Books].

Groeneveld, J., Müller, B., Angermüller, F., Drees, R., Dreßler, G., Klassert, C., et al. (2012). Good Modelling Practice: Expanding the ODD Model Description Protocol for Socioenvironmental Agent Based Models. In R. Seppelt, A. A. Voinov, S. Lange, & D. Bankamp (Eds.), *Managing Resources of a Limited Planet: Pathways and Visions Under Uncertainty. Proceedings of the Sixth Biennial Meeting of the International Environmental Modelling and Software Society,* Leipzig, Germany, July 1–5, 2012 (pp. 2480–2484). Manno: iEMSs.

Groß, M. (2001). *Die Natur der Gesellschaft. Eine Geschichte der Umweltsoziologie.* Juventa: Weinheim, München.

Groß, M., Hoffmann-Riem, H., & Krohn, W. (2005). *Realexperimente. Ökologische Gestaltungsprozesse in der Wissensgesellschaft.* Bielefeld: transcript.

Grunenberg, H. (2005). Rezension zu: Jo Reichertz (2003). Die Abduktion in der qualitativen Sozialforschung. In *Forum Qualitative Sozialforschung/ Forum: Qualitative Social Resarch, 6*(2), Art. 17. Retrieved August 2, 2005, from http://www.qualitativeresearch.net/fqs-texte/2-05/05-2-17-d.htm.

Habermas, J. (1968). *Technik und Wissenschaft als "Ideologie".* Frankfurt am Main: Suhrkamp.

Habermas, J. (1971). *Erkenntnis und Interesse.* Frankfurt am Main: Suhrkamp.

Habermas, J. (1976a [1964]). A Positivistically Bisected Rationalism. In T. W. Adorno, H. Albert, R. Dahrendorf, J. Haber, H. Pilot, & K. R. Popper

(Eds.), *The Positivist Dispute in German Sociology* (G. Adey & D. Frisby, Trans.) (pp. 198–225). London: Heinemann.

Habermas, J. (1976b). *Zur Rekonstruktion des Historischen Materialismus.* Frankfurt am Main: Suhrkamp.

Habermas, J. (1981a). *Theorie des kommunikativen Handelns. Band 1: Handlungsrationalität und gesellschaftliche Rationalisierung.* Frankfurt am Main: Suhrkamp [also published in English: Habermas, J. (1984). *Theory of Communicative Action, Volume One: Reason and the Rationalization of Society* (T. A. McCarthy, Trans.). Boston, MA: Beacon Press].

Habermas, J. (1981b). *Theorie des kommunikativen Handelns. Band 2: Zur Kritik der funktionalistischen Vernunft.* Frankfurt am Main: Suhrkamp [also published in English: Habermas, J. (1987). *Theory of Communicative Action, Volume Two: Lifeworld and System: A Critique of Functionalist Reason* (T. A. McCarthy, Trans.). Boston, MA: Beacon Press].

Habermas, J. (1991). *Erläuterungen zur Diskursethik.* Frankfurt am Main: Suhrkamp.

Habermas, J. (1995a). *Vorstudien und Ergänzungen zur Theorie des kommunikativen Handelns.* Frankfurt am Main: Suhrkamp.

Habermas, J. (1995b). Peirce and Communication. In K. L. Ketner (Ed.), *Peirce and Contemporary Thought. Philosophical Inquiries* (pp. 243–266). New York: Fordham University Press.

Habermas, J. (1999). *Wahrheit und Rechtfertigung. Philosophische Aufsätze.* Frankfurt am Main: Suhrkamp [also published in English: Habermas, J. (2003). *Truth and Justification* (B. Fultner, Ed. and Trans.). Cambridge: Polity Press].

Habermas, J. (2001). *Die Zukunft der menschlichen Natur. Auf dem Weg zur liberalen Eugenik?* Frankfurt am Main: Suhrkamp [also published in English: Habermas, J. (2003). *The Future of Human Nature.* Cambridge: Polity Press].

Habermas, J. (2008). *Between Naturalism and Religion: Philosophical Essays.* Cambridge, UK and Malden, MA: Polity Press.

Habermas, J. (2015 [1994]). *Between Facts and Norms: Contributions to a Discourse Theory of Law and Democracy.* New York, NY: Wiley.

Hämäläinen, R. P., & Lahtinen, T. J. (2016). Path Dependence in Operational Research—How the Modeling Process Can Influence the Results. *Operations Research Perspectives, 3,* 14–20.

Hampe, M. (1997). Unser Glaube an die Existenz abwesender Tatsachen. In J. Kulenkampff (Ed.), *David Hume, Eine Untersuchung über den menschlichen Verstand* (pp. 73–94). Berlin: Akademie-Verlag.

Hannigan, J. A. (1996). *Environmental Sociology. A Social Constructionist Perspective.* London: Routledge.

Hanson, N. R. (1965 [1958]). *Patterns of Discovery: An Inquiry into the Conceptual Foundations of Science.* London: Cambridge University Press.

Haraway, D. (1985). A Manifesto for Cyborgs: Science, Technology, and Socialist Feminism in the 1980s. *Socialist Review, 80,* 65–108.

Hawking, S. (2000). *Die illustrierte kurze Geschichte der Zeit.* Reinbek bei Hamburg: Rowohlt.

Heidenreich, M. (2003). Die Debatte um die Wissensgesellschaft. In S. Böschen & I. Schulz-Schaeffner (Eds.), *Wissenschaft in der Wissensgesellschaft* (pp. 25–51). Opladen: Westdeutscher Verlag.

Herrick, C., & Jamieson, D. (1995). The Social Construction of Acid Rain. *Global Environmental Change, 5*(2), 105–112.

Hirsch Hadorn, G., Hoffmann-Riem, H., Biber-Klemm, S., Grossenbacher-Mansuy, W., Joye, D., Pohl, C., et al. (Eds.). (2008). *Handbook of Transdisciplinary Research.* Berlin: Springer.

Höffe, O. (1996). *Immanuel Kant.* 4, durchges. Aufl., München: Beck.

Horkheimer, M. (1988 [1937]). Traditionelle und kritische Theorie. In M. Horkheimer & G. Schriften, *Band 4: Schriften 1936-1941* (pp. 162–216). Frankfurt am Main: Suhrkamp.

Horkheimer, M., & Adorno, T. W. (1969 [1944]). *Dialektik der Aufklärung. Philosophische Fragmente.* Frankfurt am Main: Fischer.

Huber, J. (1995). *Nachhaltige Entwicklung. Strategien für eine ökologische und soziale Erdpolitik.* Berlin: Edition Sigma.

Huber, J. (2000). Industrielle Ökologie. Konsistenz, Effizienz und Suffizienz in zyklusanalytischer Betrachtung. In R. Kreibich & U. E. Simonis (Eds.), *Global Change* (pp. 109–126). Berlin: Verlag Arno Spitz.

Huber, J. (2011). Ökologische Modernisierung und Umweltinnovation. In M. Groß (Ed.), *Handbuch Umweltsoziologie* (pp. 279–302). Wiesbaden: VS Verlag.

Hume, D. (1888 [1739/1740]). *A Treatise of Human Nature.* Reprinted from the Original Edition in Three Volumes (L. A. Selby-Bigge, Ed.). Oxford: Clarendon Press.

Hume, D. (1910 [1748]). An Enquiry Concerning Human Understanding. In C. W. Eliot (Ed.), *English Philosophers of the Seventeenth and Eighteenth Centuries: Locke, Berkeley, Hume* (Vol. 37, pp. 289–420). New York: P. F. Collier and Son.

Hurni, H., & Wiesmann, U. M. (2014). Transdisciplinarity in Practice. Experience from a Concept-Based Research Programme Addressing Global Change and Sustainable Development. *GAIA, 23*(3), 275–277.

IPCC. (1996). Climate Change 1995: The Science of Climate Change. In J. T. Houghton, L. G. M. Filho, B. A. Callander, N. Harris, A. Kattenberg, & K. Maskell (Eds.), *Contribution of Working Group I to the Second Assessment Report of the Intergovernmental Panel on Climate Change.* New York: Cambridge University Press.

IPCC. (2001). Climate Change 2001: The Scientific Basis. In J. T. Houghton, Y. Ding, D. J. Griggs, M. Noguer, P. J. van der Linden, X. Dai, K. Maskell, & C. A. Johnson (Eds.), *Contribution of Working Group I to the Third Assessment Report of the Intergovernmental Panel on Climate Change.* New York: Cambridge University Press.

IPCC. (2007). Climate Change 2007: The Physical Science Basis. In S. Solomon, D. Qin, M. Manning, Z. Chen, M. Marquis, K. B. Averyt, M. M. B. Tignor, H. L. Miller, Jr. (Eds.), *Contribution of Working Group I to the Fourth Assessment Report of the Intergovernmental Panel on Climate Change.* New York: Cambridge University Press.

Jaeger, J., & Scheringer, M. (1998). Transdisziplinarität: Problemorientierung ohne Methodenzwang. *GAIA, 7*(1), 10–25.

Janich, P. (Ed.). (1984). *Methodische Philosophie: Beiträge zum Begründungsproblem der exakten Wissenschaften in Auseinandersetzung mit Hugo Dingler.* Mannheim: B.I.-Wissenschaftsverlag.

Jetzkowitz, J. (1996). *Störungen im Gleichgewicht. Das Problem des sozialen Wandels in funktionalistischen Handlungstheorien* [Marburger Beiträge zur Sozialwissenschaftlichen Forschung; Bd. 7], Münster: LIT-Verlag.

Jetzkowitz, J. (2000). *Recht und Religion in der modernen Gesellschaft: Soziologische Theorie und Analyse am Beispiel der Rechtsprechung des Bundesverfassungsgerichts in Sachen "Religion" zwischen den Jahren 1983 und 1997.* Münster: LIT-Verlag.

Jetzkowitz, J. (2003). Funktionale Analyse als Zeichenprozess – Parsons' Soziologie als Theorie semiotischer Subjekte. In J. Jetzkowitz & C. Stark (Eds.), *Soziologischer Funktionalismus. Zur Methodologie einer Theorietradition* (141–176). Opladen: Leske+Budrich.

Jetzkowitz, J. (2008). Die Anpassung an den Klimawandel im Blick der Gesellschaftstheorie. In Institut WAR (Ed.), *Klimawandel – Markt für Strategien und Technologien?!* (pp. 99–114). Darmstadt: Schriftenreihe WAR Nr.196.

Jetzkowitz, J. (2010). "Menschheit", "Sozialität" und "Gesellschaft" als Dimensionen der Soziologie. Anregungen aus der Nachhaltigkeitsforschung. In G. Albert, R. Greshoff, & R. Schützeichel (Eds.), *Dimensionen und Konzeptionen von Sozialität.* Wiesbaden: VS Verlag für Sozialwissenschaften, 257–268.

Jetzkowitz, J. (2011). Ökosystemdienstleistungen in soziologischer Perspektive. In M. Groß (Ed.), *Handbuch Umweltsoziologie* (pp. 303–324). Wiesbaden: VS Verlag für Sozialwissenschaften.

Jetzkowitz, J., van Koppen, C. S. A. (Kris), Lidskog, R., Ott, K., Voget-Kleschin, L., & Mei Ling Wong, C. (2017). The Significance of Meaning: Why IPBES Needs the Social Sciences and Humanities. *Innovation: The European Journal of Social Science Research, 31*(suppl.1), S38–S60. https://doi.org/10.1080/13511610.2017.1348933.

Joas, H. (1992). *Die Kreativität des Handelns.* Frankfurt am Main: Suhrkamp.

Jobe, T. H. (1986). Review of "Occult and Scientific Mentalities in the Renaissance" by Brian Vickers. In *The Sixteenth Century Journal, XVII* (1), 113–115.

Jüdes, U. (1997). Nachhaltige Sprachverwirrung. Auf der Suche nach einer Theorie des Sustainable Development. *Politische Ökologie, 15*(52), 26–29.

Kamper, D. (1997). Mensch. In Christoph Wulff (Ed.), *Vom Menschen. Handbuch Historische Anthropologie* (pp. 85–91). Weinheim und Basel: Beltz.

Kampits, P. (1981). Natur als Mitwelt. Anmerkungen zu einer ökologischen Ethik. In E. Morscher & R. Stranzinger (Eds.), *Ethik. Grundlagen, Probleme und Anwendungen. Akten des 5* (pp. 328–330). Internationalen Wittgenstein Symposion (25–31 August 1980). Wien: Hölder-Pichler-Tempsky.

Kant, I. (1899 [1781/1787]). *The Critique of Pure Reason* (J. M. D. Meiklejohn, Trans.). New York, NY: Willey.

Kant, I. (1999). *Correspondence* (A. Zweig, Trans.). Cambridge: Cambridge University Press.

Kant, I. (2004 [1786]). *Metaphysical Foundations of Natural Science* (M. Friedman, Ed. and Trans.). Cambridge: Cambridge University Press.

Kapitan, T. (1994). Inwiefern sind abduktive Schlüsse kreativ? In H. Pape (Ed.), *Kreativität und Logik: Charles S. Peirce und das philosophische Problem des Neuen* (pp. 144–158). Frankfurt am Main: Suhrkamp.

Kates, R. W., Clark, W. C., Corell, R., Hall, J. M., Jaeger, C. C., Lowe, I., et al. (2001). Sustainability Science. *Science, 292*, 641–642.

Kaufmann, F.-X. (1989). *Religion und Modernität: Sozialwissenschaftliche Perspektiven.* Tübingen: J. C. B. Mohr (Paul Siebeck).

Keil, G. (1985). *Philosophiegeschichte I. Von der Antike bis zur Renaissance.* Kohlhammer: Stuttgart.

Keil, G. (1987). *Philosophiegeschichte II. Von der Aufklärung bis zur Gegenwart.* Kohlhammer: Stuttgart.

Kelle, U. (1994). *Empirisch begründete Theoriebildung. Zur Logik und Methodologie interpretativer Sozialforschung*. Weinheim: Deutscher Studien Verlag.

Kemp, R., & Martens, P. (2007). Sustainable Development: How to Manage Something That Is Subjective and Never Can Be Achieved? *Sustainability: Science, Practice, & Policy, 3*(2), 1–10.

Kizuka, T., Akasaka, M., Kadoya, T., & Takamura, N. (2014). Visibility from Roads Predict the Distribution of Invasive Fishes in Agricultural Ponds. *PLoS One, 9*(6), e99709. https://doi.org/10.1371/journal.pone.0099709.

Klintman, M. (2017). *Human Sciences and Human Interests: Integrating the Social, Economic, and Evolutionary Sciences*. London and New York: Routledge.

Knorr-Cetina, K. (1981). *The Manufacture of Knowledge: An Essay on the Constructivist and Contextual Nature of Science*. Oxford and New York: Pergamon Press.

Koestler, A. (1959). *The Sleepwalkers. A History of Man's Changing Vision of the Universe*. Macmillan: New York.

König, R. (1967). Biosoziologie. In R. König (Ed.), *Soziologie* (pp. 48–53). Frankfurt am Main: Fischer.

Kovarik, W. (2004). *Environmental History Timeline*. Retrieved November 10, 2004, from http://www.radford.edu/~wkovarik/envhist/.

Krafft, F. (1973). Johannes Keplers Beitrag zur Himmelsphysik. In F. Krafft, K. Meyer, & B. Sticker (Eds.), *Internationales Kepler Symposium, Weil der Stadt, 1971, Referate und Diskussionen* (pp. 55–139). Hildesheim: Gerstenberg.

Krafft, F. (2005). Johannes Kepler – Die neue, ursächlich begründete Astronomie. In J. Kepler, *Astronomia Nova. Neue, ursächlich begründete Astronomie* (F. Krafft, Trans.). Wiesbaden: Marix (pp. V–LVIV).

Krämer, W., & Mackenthun, G. (2001). *Die Panik-Macher*. München: Piper.

Kroll, G. (2001). The "Silent Springs" of Rachel Carson: Mass Media and the Origins of Modern Environmentalism. *Public Understanding of Science, 10*(4), 403–420.

Kropp, C. (2006). "Enacting Milk": Die Akteur-Netz-Werke von "Bio-Milch". In M. Voss & B. Peuker (Eds.), *Verschwindet die Natur? Die Akteur-Netzwerk-Theorie in der umweltsoziologischen Diskussion* (pp. 203–232). Bielefeld: transcript.

Kuhn, T. S. (1962). *The Structure of Scientific Revolutions*. Chicago: University of Chicago Press.

Kuhn, T. S. (1977). Neue Überlegungen zum Begriff des Paradigma. In T. S. Kuhn, *Die Entstehung des Neuen. Studien zur Struktur der Wissenschaftsgeschichte* (389–420). Frankfurt am Main: Suhrkamp.

Küppers, G., Lundgreen, P., & Weingart, P. (1978). *Umweltforschung – die gesteuerte Wissenschaft? Eine empirische Studie zum Verhältnis von Wissenschaftsentwicklung und Wissenschaftspolitik.* Frankfurt am Main: Suhrkamp.

Latour, B. (1993 [1991]). *We Have Never Been Modern* (C. Porter, Trans.). Cambridge, MA: Harvard University Press.

Latour, B. (1999). *Pandora's Hope: Essays on the Reality of Science Studies.* Cambridge, MA: Harvard University Press.

Latour, B. (2001). Eine Soziologie ohne Objekt? Anmerkungen zur Interobjektivität. *Berliner Journal für Soziologie, 11*(2), 237–252.

Latour, B. (2004). *Politics of Nature: How to Bring the Sciences into Democracy* (C. Porter, Trans.). Cambridge, MA: Harvard University Press.

Latour, B. (2005). *Reassembling the Social: An Introduction to Actor-Network-Theory.* Oxford: Oxford University Press.

Latour, B., & Woolgar, S. (1979). *Laboratory Life. The Social Construction of Scientific Facts.* Beverly Hills: Sage.

Le Blanc, D. (2015). *Towards Integration at last? The Sustainable Development Goals as a Network of Targets. United Nations* (DESA Working Paper No. 141). Retrieved April 16, 2018, from https://www.un.org/esa/desa/papers/2015/wp141_2015.pdf.

Lear, L. (1997). *Rachel Carson. Witness for Nature.* London: Penguin Books.

Lee, K. N. (1993). *Compass and Gyroscope: Integrating Science and Politics for the Environment.* Washington, DC: Island Press.

Lemcke, M. (2002). *Johannes Kepler.* Reinbek bei Hamburg: Rowohlt.

Lessenich, S. (2017). *Neben uns die Sintflut: Die Externalisierungsgesellschaft und ihr Preis.* Berlin: Hanser.

Liedman, S.-E. (1998). Engels and the Laws of Dialectics. In R. Panasiuk & L. Nowak (Eds.), *Marx's Theories Today* (pp. 15–35). Amsterdam: Rodopi B.V.

Lindemann, G. (2006). Die Emergenzfunktion und die konstitutive Funktion des Dritten. Perspektiven einer kritisch-systematischen Theorieentwicklung. *Zeitschrift für Soziologie, 35*(2), 82–101.

Lindemann, G. (2009). *Das Soziale von seinen Grenzen her denken.* Weilerswist: Velbrück.

Lindemann, G. (2011). Die Gesellschaftstheorie von der Sozialtheorie her denken – oder umgekehrt? *ZfS-FORUM, 3*(1), 1–19. Retrieved March 31, 2014, from http://www.zfs-online.org/index.php/forum/article/viewFile/3066/2603.

Linse, U. (1986). *Ökopax und Anarchie. Eine Geschichte der ökologischen Bewegungen in Deutschland*. München: dtv.

Lipton, P. (1991). *Inference to the Best Explanation*. London: Routledge.

Lomborg, B. (2001). *The Skeptical Environmentalist*. Cambridge and New York: Cambridge University Press.

Lucas, E. (1964). Marx' und Engels' Auseinandersetzung mit Darwin: zur Differenz zwischen Marx und Engels. *International Review of Social History, 9*, 433–469.

Luederitz, C., Schäpke, N., Wiek, A., Lang, D. J., Bergmann, M., Bos, J. J., et al. 2016. Learning Through Evaluation: A Tentative Evaluative Scheme for Sustainability Transition Experiments. *Journal of Cleaner Production*, http://dx.doi.org/10.1016/j.jclepro.2016.09.005.

Luhmann, N. (1984). *Soziale Systeme. Grundriß einer allgemeinen Theorie*. Frankfurt am Main: Suhrkamp [also published in English: Luhmann, N. (1995). *Social Systems*. Stanford: Stanford University Press].

Luhmann, N. (1988a). *Die Wirtschaft der Gesellschaft*. Frankfurt am Main: Suhrkamp.

Luhmann, N. (1988b). Warum AGIL? *Kölner Zeitschrift für Soziologie und Sozialpsychologie, 40*(1), 127–139.

Luhmann, N. (1989a). *Ecological Communication* (J. Bednarz, Trans.). Chicago: University of Chicago Press [originally published in German: Luhmann, N. (1986). *Ökologische Kommunikation. Kann die moderne Gesellschaft sich auf ökologische Gefährdungen einstellen?* Opladen: Westdeutscher Verlag].

Luhmann, N. (1989b). Politische Steuerung. Ein Diskussionsbeitrag. *Politische Vierteljahresschrift, 30*(1), 4–9.

Luhmann, N. (1990). *Die Wissenschaft der Gesellschaft*. Frankfurt am Main: Suhrkamp.

Luhmann, N. (1992). Wer kennt Wil Martens? *Kölner Zeitschrift für Soziologie und Sozialpsychologie, 44*(1), 139–142.

Luhmann, N. (1997). *Die Gesellschaft der Gesellschaft*. Frankfurt am Main: Suhrkamp [also published in English: Luhmann, N. (2012/2013). *Theory of Society*. Stanford: Stanford University Press].

Luhmann, N. (2002). *Die Politik der Gesellschaft* (A. Kieserling, Ed.). Frankfurt am Main: Suhrkamp.

Luke, T. W., & White, S. K. (1985). Critical Theory, the Informational Revolution, and an Ecological Path to Modernity. In J. Forester (Ed.), *Critical Theory and Public Life* (pp. 22–53). Cambridge, MA: MIT Press.

Lutzenhiser, L. (1993). Social and Behavioral Aspects of Energy Use. *Annual Review of Energy and the Environment, 18,* 247–289.

Martens, P. (2006). Sustainability: Science or Fiction? *Sustainability: Science, Practice and Policy, 2*(1), 1–5.

Marx, K. (1887). *Capital. A Critique of Political Economy, Volume I, Book One: The Process of Production of Capital.* Moscow: Progress Publishers.

Marx, K., & Engels, F. (1977 [1888]). *Manifesto of the Communist Party.* Moscow: Progress Publishers.

Marx, K., & Engels, F. (1978 [1850/1885]). Ansprache der Zentralbehörde an den Bund vom März 1850. In K. Marx & F. Engels, *Werke* (Band 7, pp. 244–254). Berlin: Dietz-Verlag.

Matthes, J. (1985). Die Soziologen und ihre Wirklichkeit. In W. Bonß, H. Hartmann (Eds.), *Entzauberte Wissenschaft. Sonderband 3 der Sozialen Welt* (pp. 49–64). Göttingen: Otto Schwartz und Co.

Maus, H. (1967). Zur Vorgeschichte der empirischen Sozialforschung. In R. König (Ed.), *Handbuch der empirischen Sozialforschung* (Vol. 1, pp. 21–56). Stuttgart: Ferdinand Enke.

Maxeiner, D., & Miersch, M. (1999). *Lexikon der Öko-Irrtümer.* Frankfurt am Main: Eichborn.

Mayntz, R. (Ed.). (1980). *Implementation Politischer Programme. Empirische Forschungsberichte.* Athenäum: Königstein/Ts.

Mayntz, R. (Ed.). (1983). *Implementation politischer Programme II – Ansätze zur Theoriebildung.* Opladen: Westdeutscher Verlag.

Mayntz, R. (1987). Politische Steuerung und gesellschaftliche Steuerungsprobleme. Anmerkungen zu einem theoretischen Paradigma. In T. Ellwein, J. J. Hesse, R. Mayntz, & F. W. Scharpf (Eds.), *Jahrbuch zur Staats- und Verwaltungswissenschaft* (Vol. 1, pp. 89–109). Baden-Baden: Nomos.

Mayntz, R., & Scharpf, F. W. (Eds.). (1995). *Gesellschaftliche Selbstregulierung und politische Steuerung.* Frankfurt am Main: Campus.

Mead, G. H. (1967). *Mind, Self, and Society from the Standpoint of a Social Behaviorist* (C. W. Morris, Ed., with introduction). Chicago and London: The University of Chicago Press.

Meadows, D. (2000). Es ist zu spät für eine nachhaltige Entwicklung. Nun müssen wir für eine das Überleben sichernde Entwicklung kämpfen. In W. Krull (Ed.), *Zukunftsstreit* (pp. 125–149). Weilerswrist: Velbrück.

Meadows, D. L., Meadows, D. H., Randers, J., Behrens III, W. W. (1972). *The Limits to Growth: A Report for the Club of Rome's Project on the Predicament of Mankind.* New York: Universe Books.

Meléghy, T. (2003). Methodologische Grundlagen einer evolutionären Soziologie. In T. Meléghy & H.-J. Niedenzu (Eds.), *Soziale Evolution. Die Evolutionstheorie und die Sozialwissenschaften. Sonderband 7 der Österreichischen Zeitschrift für Soziologie* (pp. 114–146). Opladen: Westdeutscher Verlag.

Merchant, C. (1980). *The Death of Nature: Women, Ecology, and the Scientific Revolution.* San Francisco: Harper & Row.

Merton, R. K. (1936). The Unanticipated Consequences of Purposive Social Action. *American Sociological Review, 1*(6), 894–904.

Michels, R. (1949). *Political Parties.* Glencoe, IL.: Free Press.

Mikkelson, G. M., Gonzales, A., & Peterson, G. D. (2007). Economic Inequality Predicts Biodiversity Loss. *PLoS ONE, 2*(5), e444. https://doi.org/10.1371/journal.pone.0000444.

Mittelstraß, J. (1992). Auf dem Wege zur Transdisziplinarität. *GAIA, 1*(5), 250.

Mittelstraß, J., Schroeder-Heister, P. (1997). Zeichen, Kalkül, Wahrscheinlichkeit. Elemente einer Mathesis universalis bei Leibniz. In H. Stachowiak (Ed.), *Pragmatik. Handbuch Pragmatisches Denken* (Vol. 1, pp. 392–414). Darmstadt: Wissenschaftliche Buchgesellschaft.

Mol, A. P. J., & Spaargaren, G. (2006). Towards a Sociology of Environmental Flows: A New Agenda for 21st Century Environmental Sociology. In G. Spargaaren, A. P. J. Mol, & F. H. Buttel (Eds.), *Governing Environmental Flows: Global Challenges to Social Theory* (pp. 39–82). Cambridge, MA: MIT Press.

Mooney, H. A., Duraiappah, A., & Larigauderie, A. (2013). Evolution of Natural and Social Science Interactions in Global Change Research Programs. *Proceedings of the National Academy of Sciences,* January 2013 (201107484). https://doi.org/10.1073/pnas.1107484110.

Moore, J. W. (2015). *Capitalism in the Web of Life: Ecology and the Accumulation of Capital.* London et al.: Verso.

Moser, H. (1978). Einige Aspekte der Aktionsforschung im internationalen Vergleich. In H. Moser & H. Ornauer (Eds.), *Internationale Aspekte der Aktionsforschung* (pp. 173–189). München: Kösel.

Müller, B., Bohn, F., Dreßler, G., Groeneveld, J., Klassert, C., Martin, R., et al. (2013). Describing Human Decisions in Agent-Based Models— Odd + D, an Extension of the Odd Protocol. *Environmental Modelling & Software, 48,* 37–48.

Münch, R., & Lahusen, C. (2001). *Democracy at Work: A Comparative Sociology of Environmental Regulation in the United Kingdom, France, Germany, and the United States.* Westport, CT: Praeger.

Murcott, S. (1997). Sustainable Development: A Meta-Review of Definitions, Principles, Criteria, Indicators, Conceptual Frameworks, Information Systems. In *Annual Conference of the American Association for the Advancement of Science. IIASA Symposium on "Sustainability Indicators".* Seattle, WA, February 13–18, 1997.

Naess, A. (1973). The Shallow and the Deep, Long-Range Ecology Movement: A Summary. *Inquiry, 16*(1), 95–100.

Nerlich, B. (2003). Tracking the Fate of the Metaphor *Silent Spring* in British Environmental Discourse. Towards an Evolutionary Ecology of Metaphor. In *Metaphoric* 04/2003, 115–140.

Neurath, O. (1932/1933). Protokollsätze. *Erkenntnis, 3*(1), 204–214 [English Translation in: R. S. Cohen & M. Neurath (Eds.), *Philosophical Papers 1913–1946. Vienna Circle Collection* (Vol. 16, pp. 91–99). Dordrecht: Springer].

Nicolescu, B. (2005). *Transdisciplinarity—Past, Present and Future.* Retrieved March 28, 2011, from http://cetrans.com.br/textos/transdisciplinarity-past-present-and-future.pdf.

Nicolescu, B. (2007). *Transdisciplinarity as Methodological Framework for Going Beyond the Science-Religion Debate.* Retrieved March 28, 2011, from http://www.fernandosantiago.com.br/transdisx.pdf.

Nilsson, M., Griggs, D., & Visbeck, M. (2016). Map the Interactions Between Sustainable Development Goals. *Nature, 534,* 320–322.

Norgaard, R. B. (1994). *Development Betrayed: The End of Progress and a Coevolutionary Revisioning of the Future.* London: Routledge.

Norton, B. G. (2005). *Sustainability. A Philosophy of Adaptive Ecosystem Management.* Chicago: The University of Chicago Press.

Nowotny, H. (2001). Vom Geschichtenerzählen zur Koevolutionswissenschaft. *GAIA, 10*(4), 262–264.

Nowotny, H., Scott, P., & Gibbons, M. (2001). Re-Thinking Science: Knowledge and the Public in an Age of Uncertainty. Cambridge: Polity.

Oehler, K. (1995). A Response to Habermas. In K. L. Ketner (Ed.), *Peirce and Contemporary Thought. Philosophical Inquiries* (pp. 267–271). New York: Fordham University Press.

Oevermann, U. (1991). Genetischer Strukturalismus und das sozialwissenschaftliche Problem der Erklärung der Entstehung des Neuen. In S. Müller-Doohm (Ed.), *Jenseits der Utopie. Theoriekritik der Gegenwart* (pp. 267–336). Frankfurt am Main: Suhrkamp.

Ostrom, E. (2009). A General Framework for Analyzing Sustainability of Social-Ecological Systems. *Science, 325,* 419–422.

Ott, K. (1993). *Ökologie und Ethik: Ein Versuch praktischer Philosophie.* Tübingen: Attempto-Verlag.

Ott, K. (2010). *Umweltethik zur Einführung.* Marburg: Metropolis-Verlag.

Ott, K., & Döring, R. (2008). *Theorie und Praxis starker Nachhaltigkeit.* Marburg: Metropolis-Verlag.

Outram, D. (1995). *The Enlightenment.* Cambridge: Cambridge University Press.

Pahl-Wostl, C., Giupponi, C., Richards, K., Binder, C., de Sherbinin, A., Sprinz, D., et al. (2013). Transition Towards a New Global Change Science: Requirements for Methodologies, Methods, Data and Knowledge. *Environmental Science & Policy, 28,* 36–47.

Pape, H. (1989). *Erfahrung und Wirklichkeit als Zeichenprozeß. Charles S. Peirces Entwurf einer Spekulativen Grammatik des Seins.* Suhrkamp: Frankfurt am Main.

Pape, H. (1991). Einleitung. In C. S. Peirce (Ed.), *Naturordnung und Zeichenprozeß. Schriften über Semiotik und Naturphilosophie* (pp. 11–109). Frankfurt am Main: Suhrkamp.

Pape, H. (1994). Zur Einführung: Logische und metaphysische Aspekte einer Philosophie der Kreativität. C. S. Peirce als Beispiel. In H. Pape (Ed.), *Kreativität und Logik: Charles S. Peirce und das philosophische Problem des Neuen* (pp. 9–59). Frankfurt am Main: Suhrkamp.

Pape, H. (1995). *The Social Nature of Reality and Communication: Peirce vs. Mead?* Unpublished Manuscript.

Pape, H. (1999). *Abduction and the Typology of Human Cognition. Transactions of the Charles S. Peirce Society, 35*(2), 248–269.

Pape, H. (2002). Indexikalität und die Anwesenheit der Welt in der Sprache. In M. Kettner & H. Pape (Eds.), *Indexikalität und sprachlicher Weltbezug* (pp. 91–119). Paderborn: Mentis.

Parsons, T. (1964). A Functional Theory of Change. In A. Etzioni & E. Etzioni (Eds.), *Social Change: Sources, Patterns, and Consequences* (pp. 83–97). New York: Basic Books.

Parsons, T. (1966). *Societies: Evolutionary and Comparative Perspectives.* Englewood Cliffs, NJ: Prentice Hall.

Parsons, T. (1971). *The System of Modern Societies.* Englewood Cliffs, NJ: Prentice Hall.

Parsons, T. (1977). Comparative Studies and Evolutionary Change. In T. Parsons (Ed.), *Social Systems and the Evolution of Action Theory* (pp. 279–320). New York and London: The Free Press.

Peirce, C. S. (1931). The Classification of Sciences. In, C. Hartshorne & P. Weiss (Eds.), *Collected Papers, Vol. I: Principles of Philosophy* (pp. 75–137). Cambridge, MA: Harvard University Press.

Peirce, C. S. (1934a). Lectures on Pragmatism. In C. Hartshorne & P. Weiss (Eds.), *Collected Papers, Vol. V: Pragmatism and Pragmaticism* (pp. 13–131). Cambridge, MA: Harvard University Press.

Peirce, C. S. (1934b). Some Consequences of Four Incapabilities. In C. Hartshorne & P. Weiss (Eds.), *Collected Papers, Vol. V: Pragmatism and Pragmaticism* (pp. 156–189). Cambridge, MA: Harvard University Press.

Peirce, C. S. (1934c). The Fixation of Belief. In C. Hartshorne & P. Weiss (Eds.), *Collected Papers, Vol. V: Pragmatism and Pragmaticism* (pp. 293–313). Cambridge, MA: Harvard University Press.

Peirce, C. S. (1934d). The Fixation of Belief. In C. Hartshorne & P. Weiss (Ed.), *Collected Papers, Vol. V: Pragmatism and Pragmaticism* (pp. 293–313). Cambridge, MA: Harvard University Press.

Peirce, C. S. (1958). *Collected Papers, Vol. VII: Science and Philosophy* (A. W. Burks, Ed.). Cambridge, MA: Harvard University Press.

Peirce, C. S. (1976). *The New Elements of Mathematics, Vol 4: Mathematical Philosophy* (C. Eisele, Ed.). The Hague: Mouton Publishers.

Peirce, C. S. (1983 [1903]). *Phänomen und Logik der Zeichen* (H. Pape, Ed. and Trans.). Frankfurt am Main: Suhrkamp.

Peirce, C. S. (1986 [1901]). Minutiöse Logik. Aus den Entwürfen zu einer Logik. In C. S. Peirce, *Semiotische Schriften* (Vol. I, Ed. and Trans. C. Kloesel & H. Pape, pp. 376–408). Frankfurt am Main: Suhrkamp.

Peirce, C. S. (1991). *Naturordnung und Zeichenprozeß. Schriften über Semiotik und Naturphilosophie* (H. Pape, Ed. and Intro.). Frankfurt am Main: Suhrkamp.

Peirce, C. S. (1998). *The Essential Peirce: Selected Philosophical Writings, Vol. 2 (1893–1913)* (N. Houser et al., Ed.). Bloomington: Indiana University Press.

Phillips, D. (2003). *The Truth of Ecology. Nature, Culture, and Literature in America.* Oxford: Oxford University Press.

Pilot, H. (1972). *Prolegomena zu einer kritischen Theorie der Erfahrung.* Heidelberg: Dissertation Philosophisch-historische Fakultät.

Pindyck, R. S. (2017). The Use and Misuse of Models for Climate Policy. *Review of Environmental Economics and Policy, 11*(1), 100–114.

Platon. (1990a). Menon.In *Platon, Werke in acht Bänden. Griechisch – Deutsch* (Vol. 2, G. Eigler, Ed., pp. 505–599). Darmstadt: Wissenschaftliche Buchgesellschaft.

Platon. (1990b). Politeia. In *Platon, Werke in acht Bänden. Griechisch – Deutsch* (Vol. 2, G. Eigler, Ed., pp. 505–599). Darmstadt: Wissenschaftliche Buchgesellschaft.

Popper, K. R. (1950 [1945]). *The Open Society and Its Enemies.* Princeton, NJ: Princeton University Press.

Popper, K. R. (1963). *Conjectures and Refutations: The Growth of Scientific Knowledge.* London: Routledge and Kegan Paul.

Popper, K. R. (1979 [1972]). *Objective Knowledge: An Evolutionary Approach.* Oxford: Clarendon Press.

Popper, K. R. (1994[1934]). *Logik der Forschung.* Tübingen: Mohr [also published in English: Popper, K. R. (1965). *The Logic of Scientific Discovery.* New York: Harper & Row].

Porter, R. (2001). *The Enlightenment.* Basingstoke: Palgrave.

Portmann, A. (1972). Tiersoziologie. In W. Bernsdorf (Ed.), *Wörterbuch der Soziologie 3* (pp. 857–861). Frankfurt am Main: Fischer.

Pradhan, P., Costa, L., Rybski, D., Lucht, W., & Kropp, J. P. (2017). A Systematic Study of Sustainable Development Goal (SDG) Interactions. *Earth's Future, 5,* 1169–1179.

Pregernig, M., Rhodius, R., & Winkel, G. (2018). Design Junctions in Real-World Laboratories: Analyzing Experiences Gained from the Project Knowledge Dialogue Northern Black Forest. *GAIA, 27*(S1), 32–38.

Prigogine, I. (1980). *From Being to Becoming: Time and Complexity in the Physical Sciences.* San Francisco: Freeman.

Prigogine, I., & Stengers, I. (1984). *Order out of Chaos: Man's new Dialogue with Nature.* New York: Bantam Books.

ProClim/CASS, Konferenz der Schweizerischen Wissenschaftlichen Akademien (Eds.). (1997). *Forschung zu Nachhaltigkeit und globalem Wandel – wissenschaftspolitische Visionen der schweizer Forschenden.* Bern: ProClim.

Radkau, J. (2008). *Nature and Power: A Global History of the Environment* (T. Dunlap, Trans.). Cambridge: Cambridge University Press.

Radkau, J. (2011). *Die Ära der Ökologie. Eine Weltgeschichte.* München: Beck [also published in English: Radkau, J. (2014). *The Age of Ecology: A Global History.* Cambridge: Polity Press].

Radnitzky, G. (1989). Wissenschaftstheorie, Methodologie. In H. Seiffert & G. Radnitzky (Eds.), *Handlexikon zur Wissenschaftstheorie* (pp. 463–471). München: Ehrenwirth.

Raeithel, A. (1994). Symbolic Production of Social Coherence. The Evolution of Dramatic, Discursive and Objectified Meaning Systems. *Mind, Culture and Activity, 1*(1–2), 69–123.

Rammert, W. (1997). New Rules of Sociological Method: Rethinking Technology Studies. *British Journal of Sociology, 48*(2), 171–191.

Rantis, K. (2004). *Geist und Natur. Von den Vorsokratikern zur Kritischen Theorie.* Darmstadt: Wissenschaftliche Buchgesellschaft.

Reichenbach, H. (1938). *Experience and Prediction: An Analysis of the Foundations and the Structure of Knowledge.* Chicago: University of Chicago Press.

Reichertz, J. (1991). *Aufklärungsarbeit. Kriminalpolizisten und Feldforscher bei der Arbeit.* Stuttgart: Ferdinand Enke.

Reichertz, J. (2003). *Die Abduktion in der qualitativen Sozialforschung.* Opladen: Leske+Budrich.

Reichholf, J. H. (2002). *Die falschen Propheten. Unsere Lust an Katastrophen.* Berlin: Klaus Wagenbach Verlag.

Renn, O. (1985). Die alternative Bewegung: Eine historisch-soziologische Analyse des Protestes gegen die Industriegesellschaft. *Zeitschrift für Politik, 32,* 152–194.

Ridley, M. (2003). *Nature via Nurture: Genes, Experience, and What Makes Us Human.* New York: Harper Collins.

Rip, A. (2002). Co-Evolution of Science, Technology and Society. An Expert Review for the Bundesministerium Bildung und Forschung's Förderinitiative Politik, Wissenschaft und Gesellschaft (Science Policy Studies), as managed by the Berlin-Brandenburgische Akademie der Wissenschaften. Enschede: University of Twente, June 2002. Retrieved March 19, 2014, from http://citeseerx.ist.psu.edu/viewdoc/download?doi=10.1.1.201.6112&rep=rep1&type=pdf.

Rogers, E. M. (1962). *Diffusion of Innovations.* New York: The Free Press of Glencoe.

Røpke, I. (2004). The Early History of Modern Ecological Economics. *Ecological Economics, 50*(3/4), 293–314.

Røpke, I. (2005). Trends in the Development of Ecological Economics from the Late 1980s to the Early 2000s. *Ecological Economics, 55*(2), 262–290.

Sachs, W. (1983). The Social Construction of Energy: A Chapter in the History of Scarcity. In *Schriftenreihe "Energie und Gesellschaft"*, Heft 22. Berlin: Technische Universität Berlin.

Sandker, M., Campbell, B. M., Ruiz-Pérez, M., Sayer, J. A., Cowling, R., Kassa, H., & Knight, A. T. (2010). The Role of Participatory Modeling in Landscape Approaches to Reconcile Conservation and Development. *Ecology and Society, 15*(2), 13. Retrieved April 20, 2018, from http://www.ecologyandsociety.org/vol15/iss2/art13/.

Schäpke, N., Stelzer, F., Caniglia, G., Bergmann, M., Wanner, M., Singer-Brodowski, M., et al. (2018). Jointly Experimenting for Transformation? Shaping Real-World Laboratories by Comparing Them. *GAIA, 27*(S1), 85–96.

Schäpke, N., Stelzer, F., Marg, O., Bergmann, M., Miller, E., Wagner, F., et al. (2017). Urban BaWü-Labs: Challenges and Solutions When Expanding the Real-World Lab Infrastructure. *GAIA, 26*(4), 366–368.

Scharpf, F. W. (1989). Politische Steuerung und politische Institutionen. *Politische Vierteljahresschrift, 30*(1), 10–21.

Scheich, E. (1993). *Naturbeherrschung und Weiblichkeit. Denkformen und Phantasmen der modernen Naturwissenschaften.* Pfaffenweiler: Centaurus.

Schellnhuber, H.-J. (2001). Die Koevolution von Natur, Gesellschaft und Wissenschaft – Eine Dreiecksbeziehung wird kritisch. *GAIA, 10*(4), 258–262.

Scheunemann, E. (2008). *Vom Denken der Natur. Natur und Gesellschaft bei Habermas.* Norderstedt: Books on Demand.

Schnaiberg, A. (2005). The Economy and the Environment. In N. J. Smelser & R. Swedberg (Eds.), *The Handbook of Economic Sociology* (2nd ed., pp. 703–725). Princeton: Princeton University Press.

Schneider, S. H., & Londer, R. (1984). *Coevolution of Climate and Life.* San Francisco: Sierra Club Books.

Schneidewind, U., Augenstein, K., Stelzer, F., & Wanner, M. (2018). Structure Matters: Real-World Laboratories as a New Type of Large-Scale Research Infrastructure—A Framework Inspired by Giddens' Structuration Theory. *GAIA, 27*(S1), 12–17.

Schneidewind, U., Singer-Brodowski, M., Augenstein, K. (2016). Transformative Science for Sustainability Transitions. In H. G. Brauch, U. Oswald Spring, J. Grin, & J. Scheffran (Eds.), *Handbook on Sustainability Transition and Sustainable Peace* (pp. 123–136). Berlin and Heidelberg: Springer.

Schröer, H. (1990). Art. Kybernetik. *Theologische Realenzyklopädie, 20,* 356–359.

Schumpeter, J. A. (1912). *Theorie der wirtschaftlichen Entwicklung.* Leipzig: Duncker & Humblot.

Schwinn, T. (2003). Makrosoziologie jenseits von Gesellschaftstheorie. Funktionalismuskritik nach Max Weber. In J. Jetzkowitz & C. Stark (Eds.), *Soziologischer Funktionalismus. Zur Methodologie einer Theorietradition* (pp. 83–109). Opladen: Leske+Budrich.

Searle, J. R. (1983). *Intentionality: An Essay in the Philosophy of Mind.* Cambridge: Cambridge University Press.

Sellars, W. (1997). *Empiricism and the Philosophy of Mind* (R. Rorty, With an introduction, a study guide by R. R. Brandom). Cambridge, MA: Harvard University Press.

Sen, A. (1987). *On Ethics and Economics.* Oxford: Basil Blackwell.

Shove, E. (2010). Beyond the ABC: Climate Change Policy and Theories of Social Change. *Environment and Planning A, 42*(6), 1273–1285.

Siebenhüner, B., & Heinrichs, H. (2010). Knowledge and Social Learning for Sustainable Development. In M. Gross & H. Heinrichs (Eds.), *Environmental Sociology: European Perspectives and Interdisciplinary Challenges* (pp. 185–199). Doderecht: Springer.

Sieferle, R. P. (1984). *Fortschrittsfeinde? Opposition gegen Technik und Industrie von der Romantik bis zur Gegenwart.* München: Beck.

Simmel, G. (2004 [1900]). *The Philosophy of Money* (Edieted by David Frisby. Translated by Tom Bottomore and David Frisby from a First Draft by Kaethe Mengelberg). London and New York: Routledge.

Simon, H. A. (1973). Does Scientific Discovery Have a Logic? *Philosophy of Science, 40,* 471–480.

Sitter-Liver, B. (2000). Tiefen-Ökologie: Kontrapunkt im aktuellen Kulturgeschehen. *Natur und Kultur, 1*(1), 70–88.

Smith, T. M., & Reynolds, R. W. (2005). A Global Merged Land-Air-Sea Surface Temperature Reconstruction Based on Historical Observations (1880–1997). *Journal of Climate, 18*(12), 2021–2036.

Spaemann, R. (1980). Technische Eingriffe in die Natur als Problem der politischen Ethik. In D. Birnbacher (Ed.), *Ökologie und Ethik* (pp. 180–206). Stuttgart: Reclam.

Spangenberg, J. (2015). Sustainability and the Challenge of Complex Systems. In J. C. Enders, & M. Remig (Eds.), *Theories of Sustainable Development* (pp. 89–111). London and New York: Routledge.

Stachowiak, H. (1973). *Allgemeine Modelltheorie.* Wien and New York: Springer.

Stark, C. (1998). *Die blockierte Demokratie. Kulturelle Grenzen der Politik im deutschen Immissionsschutz.* Nomos: Baden-Baden.

Stark, C. (2003). Neopositivistische Gesellschaftstheorie. Ein Essay vom, Ende der Geschichte' und zur, natürlichen Ordnung' des Funktionalismus. In J. Jetzkowitz, C. Stark (Eds.), *Soziologischer Funktionalismus. Zur Methodologie einer Theorietradition* (pp. 219–246). Opladen: Leske+Budrich.

Stauffacher, M. (2011). Umweltsoziologie und Transdisziplinarität. In M. Groß (Ed.), *Handbuch Umweltsoziologie* (pp. 259–276). Wiesabaden: VS Verlag.

Stegmüller, W. (1989). *Hauptströmungen der Gegenwartsphilosophie. Eine kritische Einführung.* Band 1. Stuttgart: Alfred Kröner.

Stehr, N. (1994). *Knowledge Societies.* London: Sage.

Stehr, N., & von Storch, H. (1999). *Klima – Wetter – Mensch.* München: Beck.

Stephenson, B. (1987). *Kepler's Physical Astronomy.* New York: Springer.

Strohschneider, P. (2014). Zur Politik der Transformativen Wissenschaft. In A. Brodocz, D. Herrmann, R. Schmidt, D. Schulz, & J. Schulze-Wessel (Eds.), *Die Verfassung des Politischen. Festschrift für Hans Vorländer* (pp. 171–192). Wiesbaden: Springer.

Tembrock, G. (1997). Grundlagen und Probleme einer allgemeinen Tiersoziologie. *Ethik und Sozialwissenschaften, 8*(1), 71–80.

Tenbruck, F. H. (1981). Emile Durkheim oder die Geburt der Gesellschaft aus dem Geist der Soziologie. *Zeitschrift für Soziologie, 10*(4), 333–350.

Tenbruck, F. H. (1989). Gesellschaftsgeschichte oder Weltgeschichte? *Kölner Zeitschrift für Soziologie und Sozialpsychologie, 41*(3), 417–439.

Tester, K. (1991). *Animals and Society. The Humanity of Animal Rights.* London: Routledge.

Thagard, P. R. (1988). *Computational Philosophy of Science.* Cambridge, MA: London: The MIT Press.

Theobald, W. (2003). *Mythos Natur. Die geistigen Grundlagen der Umweltbewegung.* Darmstadt: Wissenschaftliche Buchgesellschaft.

Thomas, W. I., & Thomas, D. S. (1928). *The Child in America: Behavior Problems and Programs.* New York: Knopf.

Tiner, J. H. (1977). *Johannes Kepler. Giant of Faith and Science.* Mott Media: Milford, MI.

Touraine, A. (1986). Krise und Wandel des sozialen Denkens. In J. Berger (Ed.), *Die Moderne – Kontinuitäten und Zäsuren* (pp. 15–39). Göttingen: Otto Schwartz und Co.

Tovey, H. (2003). Theorising Nature and Society in Sociology: The Invisibility of Animals. *Sociologia Ruralis, 43*(3), 196–215.

Ulrich, P. (1986). *Transformationen der ökonomischen Vernunft. Fortschrittsperspektiven der modernen Industriegesellschaft.* Bern: Paul Haupt.

Ulrich, P. (2008). Integrative Wirtschaftsethik. Grundlagen einer lebensdienlichen Ökonomie. 4., vollständig neu berarb. Aufl. Bern: Haupt [also published in English: Ulrich, P. (2008). *Integrative Economic Ethics: Foundations of a Civilized Market Economy.* Cambridge: Cambridge University Press].

United Nations Educational, Scientific and Cultural Organization (UNESCO). (2005). Towards Knowledge Societies (UNESCO World Report). Paris: UNESCO Publishing, http://unesdoc.unesco.org/images/0014/001418/141843e.pdf.

van Eeten, M. J. G., Loucks, D. P., & Roe, E. (2002). Bringing Actors Together Around Large-Scale Water Systems: Participatory Modeling and Other Innovations. *Knowledge, Technology, & Policy, 14*(4), 94–108.

van Orman Quine,W. (1994). *From Stimulus to Science.* Cambridge, MA: Harvard University Press.

Vahrenholt, F., & Lüning, S. (2012). *Die kalte Sonne. Warum die Klimakatastrophe nicht stattfindet.* Hamburg: Hoffmann und Campe.

Vaihinger, H. (1970 [1881/1892]). *Kommentar zu Kants Kritik der reinen Vernunft* (2 vols.). Reprint of the second edition (1922) (R. Schmidt, Ed.). Aalen: Scientia.

Vickers, B. (Ed.). (1984). *Occult and Scientific Mentalities in the Renaissance.* London: Cambridge University Press.

Viehöver, W. (2003). Die Klimakatastrophe als ein Mythos der reflexiven Moderne. In L. Clausen, E. M. Geenen, & E. Macamo (Eds.), *Entsetzliche soziale Prozesse. Theorie und Empirie der Katastrophen* (pp. 247–286). Münster: Lit.

Voelkel, J. R. (2001). *The Composition of Kepler's Astronomia Nova.* Princeton and Oxford: Princeton University Press.

Voinov, A., Seppelt, R., Reis, S., Nabel, J. E. M. S., & Shokravi, S. (2014). Values in Socio-Environmental Modelling: Persuasion for Action or Excuse for Inaction. *Environmental Modelling & Software, 53,* 207–212.

von Carlowitz, H. C. (2000 [1713]). *Sylvicultura oeconomica oder haußwirthschaftliche Nachricht und naturmäßige Anweisung zur wilden Baum-Zucht.* Freiberg: TU Bergakademie Freiberg.

von Weizsäcker, C. F. (1971). *Die Einheit der Natur. Studien.* München: Carl Hanser [also published in English: Weizsäcker. (1980). *The Unity of Nature* (F. J. Zucker, Trans.). New York: Farrar, Straus, Giroux].

von Weizsäcker, E. U. (1994). Erdpolitik. Ökologische Realpolitik an der Schwelle zum Jahrhundert der Umwelt. 4., aktualisierte Aufl., Darmstadt: Wissenschaftliche Buchgesellschaft [also published in English: von Weizsäcker, E. U. (1994). *Earth Politics.* London and Atlantic Highlands, NJ: Zed Books].

Voß, J.-P., Newig, J., Kastens, B., Monstadt, J., & Nöltig, B. (2007). Steering for Sustainable Development: A Typology of Problems and Strategies with Respect to Ambivalence, Uncertainty and Distributive Power. *Journal of Environmental Policy and Planning, 9*(3/4), 193–212.

Wallerstein, I. M. (1974–1989). *The Modern World-System* (3 vols.). New York, London and San Diego, CA: Academic Press.

Wals, A. E. J. (2007). *Social Learning Towards a Sustainable World: Principles, Perspectives, and Praxis.* Wageningen: Wageningen Academic Publishers.

Weber, M. (1980 [1921]). *Wirtschaft und Gesellschaft. Grundriss der verstehenden Soziologie.* Tübingen: Mohr.

Weber, M. (2002 [1905]). *The Protestant Ethic and the Spirit of Capitalism* (S. Kahlberg, Intr. and Trans.). Oxford: Blackwell.

Whitehead, A. N. (1925). *Science and the modern world.* New York: Macmillan.

Whitehead, A. N. (1979 [1929]). *Prozeß und Realität. Entwurf einer Kosmologie. Übers. u. mit einem Nachwort v. Hans Günter Holl.* Frankfurt am Main: Suhrkamp.

White, L. T. (1967). The Historical Roots of Our Ecologic Crisis. *Science, 155*(3767), 1203–1207.

Wiesmann, U., Hurni, H., Ott, C., & Zingerli, C. (2011). Combining the Concepts of Transdisciplinarity and Partnership in Research for Sustainable Development. In U. Wiesmann & H. Hurni (Eds.), *Research for Sustainable Development: Foundations, Experiences, and Perspectives. Perspectives of the Swiss National Centre of Competence in Research (NCCR) North-South* (pp. 43–70). Bern: Geographica Bernensia.

Wilhite, H. L., & Nørgaard, J. (2004). Equating Efficiency with Reduction: A Self Deception in Energy Policy. *Energy and Environment, 15*(6), 991–1009.

Williams, E. E. (1944). *Capitalism and Slavery.* Chapel Hill: University of North Carolina Press.

Windelband, W. (1904). *Geschichte und Naturwissenschaft* (3rd ed.). Straßburg: Heitz.

Witzany, G. (2000). *Life: The Communicative Structure.* Norderstedt: Libri Books on Demand.

World Commission on Environment and Development (WCED). (1987). Report of the World Commission on Environment and Development: Our Common Future, Chapter 2: Towards Sustainable Development. Retrieved August 26, 2010, from http://www.un-documents.net/ocf-02.htm.

Worster, D. (1993). *The Wealth of Nature: Environmental History and the Ecological Imagination.* New York: Oxford University Press.

Zachos, J., Pagani, M., Sloan, L., Thomas, E., & Billups, K. (2001). Trends, Rhythms, and Aberrations in Global Climate 65 Ma to Present. *Science, 292*(5517), 686–693.

Zierhofer, W. (1994). Ist die kommunikative Vernunft der ökologischen Krise gewachsen? Ein Evaluationsversuch. In W. Zierhofer & D. Steiner (Eds.), *Vernunft angesichts der Umweltzerstörung* (pp. 161–194). Opladen: Westdeutscher Verlag.

Zilsel, E. (1976). *Die sozialen Ursprünge der neuzeitlichen Wissenschaft. Edited by Wolfgang Krohn.* Frankfurt am Main: Suhrkamp.

Zimmerli, W. C. (1997). Zeit als Zukunft. In A. Gimmler, M. Sandbothe, & W. C. Zimmerli (Eds.), *Die Wiederentdeckung der Zeit. Reflexionen, Analysen, Konzepte* (pp. 126–147). Darmstadt: Wissenschaftliche Buchgesellschaft.

Index

229

Printed by Printforce, the Netherlands